Wild Ireland

CARSTEN KRIEGER is a photographer, author and environmentalist based on the west coast of Ireland. He has published numerous books on Ireland's landscape, nature and heritage including *Ireland's Coast*, *The River Shannon* and the popular *Ireland's Wild Atlantic Way*. Carsten is also working as editor for the Crossbill Guides Foundation and project manager for AstonECO. He is an AnTaisce Climate Ambassador and a Green Sod Ireland Biodiversity Ambassador. carstenkrieger.com

For Michael O'Brien

I hope they have books where you are

Wild

A NATURE JOURNEY

Ireland

CARSTEN KRIEGER

THE O'BRIEN PRESS
DUBLIN

First published 2023 by The O'Brien Press Ltd.,
12 Terenure Road East, Rathgar, Dublin 6, D06 HD27, Ireland.
Tel: +353 1 4923333. Fax: +353 1 4922777
Email: books@obrien.ie. Website: obrien.ie
The O'Brien Press is a member of Publishing Ireland.

ISBN 978-1-78849-317-8

Quote on p9 by Baba Dioum from a paper presented in New Delhi, India,
1968, at the triennial meeting of the General Assembly of the International
Union for the Conservation of Nature and Natural Resources (IUCN). iucn.org

1 3 5 7 9 10 8 6 4 2
23 25 27 26 24

Printed by EDELVIVES, Spain.
The paper in this book is produced using pulp from managed forests.

The author is a member of and
adheres to the principles of nature and
wildlife photography as outlined by

Published in:

DUBLIN
UNESCO
City of Literature

Enjoying life with
O'BRIEN
obrien.ie

NATURE FIRST
The Alliance for Responsible Nature Photography

TABLE OF CONTENTS

Introduction

I have been mesmerised by the natural world since my early childhood in Germany. I remember sitting by the kitchen window on cold winter mornings, watching the birds at the feeder outside, and chasing butterflies around the garden on hot summer evenings. I devoured books like Bernhard Grzimek's *Serengeti darf nicht sterben* (*Serengeti Shall Not Die*) and was glued to the TV every time *Expeditionen ins Tierreich* with Heinz Sielmann – who was the German equivalent to David Attenborough – came on.

Part of this early interest was simple childish curiosity: Why is the goldfinch so colourful? Why does the caterpillar turn into a butterfly? How does the bee get the nectar out of the flower (and how does the nectar get into the flower in the first place)? Another part was the fact that I felt very comfortable in the company of wild things ... more comfortable than with most humans.

A few years later, an obsession with photography came into the mix, and soon after

Opposite: Red squirrel, County Laois.
Below: Common poppy, County Meath.

Top left: Shore crab, County Galway.
Top right: Fin whale, County Kerry.
Above left: Blue tit, County Clare.
Above right: Thrift and bumblebee, County Clare.

it was clear to me that I wanted to spend my life as a photographer and work in nature conservation.

But life had other plans. I was a lazy student and pretty much flunked all subjects apart from biology. After leaving school, I took some detours and ended up training and subsequently working as a paediatric nurse.

Some dreams, however, die hard, and after a decade in nursing, I moved to Ireland for a new beginning and to pursue photography and nature conservation as a career. By pure chance, I ended up living on the Loop Head Peninsula in County Clare, which turned out to be a treasure chest for someone like me: bottlenose dolphins, seabird colonies and rocky shores teeming with all kinds of wildlife were on my doorstep, waiting to be explored. On top of that, the Burren — a landscape I started exploring on my holiday trips to Ireland and had very much fallen in love with — was only a short drive away.

Unfortunately for me, it was difficult to make a living from nature and conservation photography alone, so I started shooting pretty landscapes and happy people for the tourism sector. This became my day job, and I worked mostly as a volunteer for various conservation groups on the side. But a change was coming. The terms 'climate change', 'biodiversity loss' and 'mass extinction' entered the public domain, and more and more people started to show an interest in the natural world and the threats it is facing.

David Attenborough summed it up when he said, 'We moved from being a part of nature to being apart from nature.' We are now at a point in history where we must re-evaluate and change our perception and relationship with the natural world. We are beginning to understand that we are merely one part of a vast global network that connects all living beings — and that, within this network, actions (our actions) have consequences.

This new understanding of our place in the structure of the planet should begin at home, on the small grassy patch outside our front door, the pasture down the road, the stretch of coast or the forest where we take our daily walks. In these places, we can experience the workings of the natural network at first hand.

Baba Dioum, a Senegalese forestry engineer, wrote in 1968, 'In the end we will conserve only what we love, we will love only what we understand, and we will understand only what we are taught', and his words are truer now than ever. I strongly believe that education is the key to bringing us back to a state where we are a part of nature again.

I don't consider myself an expert, but I have been exploring and learning about the natural world for most of my life, and I hope that this book will help you to understand, love and conserve our wild Ireland.

Below: Raven, County Clare.

Chapter I

Limestone Country

A thick blanket of fog covers the wintry landscape. The path, carved by countless feet into the shallow layer of dirt that covers the limestone, is barely visible. The damp air is swirling around me, at times lifting for the blink of an eye to allow a brief glance of wet limestone slabs and the vague outline of shrubs. Suddenly the lake emerges out of the grey wall of mist, its surface motionless, its extent hidden. It is winter solstice day, and even though it is almost lunchtime, it feels like early morning. I am on my way to the summit of Mullach Mór, the iconic flat-topped hill that sits in the heart of the Burren National Park, a limestone karst area that has mesmerised visitors and locals for millennia.

I am losing sight of the lake, but the stile in a dry-stone wall shows me that I am on the right track. A few steps further and another waymark, a small stand of hazel, becomes just

Opposite: Rine Peninsula and Ballyvaughan Bay, the Burren, County Clare.
Below: Limestone pavement at Rock Forest.

about visible to my left. Soon after, a number of natural steps in the limestone bring me onto the first plateau of Mullach Mór.

Any landmarks and way markers are now completely hidden in the fog, which is thicker than ever. The path has disappeared, and the limestone at my feet doesn't keep any memories of previous travellers. I am stranded, and all I can do is sit and wait. Time goes by, the cold and damp slowly making its way through layers of clothing.

Then, all of a sudden, it seems to be getting brighter. I can see a bit further; the curtain is lifting enough to wander on. Patches of blue appear in the sky above me, and the mountainscape of the Burren is revealing itself all around: Poulnalour and Slievenaglasha to my left, Slieve Roe and Knockanes ahead of me, and the lowlands of Clare and Galway to my right. The skeletons of the hazel scrubs glow in gentle brown and yellow tones, the wet limestone is glistening in the light, and the only sound is the hushing of the wind. This is the Burren.

❋ ❋ ❋

Most of Ireland's bedrock is made from limestone, but only in a few places has this rock come to the surface to become the landscape. It does so in counties Fermanagh and Cavan and a few odd spots in the midlands, but Ireland's ultimate limestone landscape, the Burren, lies in the northern half of County Clare and spills over into southern County Galway. The first impression of this area is one of gloomy desolation. From a distance, the first features to catch the eye are terraced, raggy and seemingly bare hills. A closer look reveals polished rock pavements that are traversed by deep fissures and strewn with rocks known as erratics.

On dry, sunny days, the rock presents itself in a bright, almost blinding, grey. After a downpour, the landscape changes to a dark grey, almost black, and glimmers with a blueish tint. And on those soft days that are so typical for Ireland, when a constant fine drizzle drenches the land, it is almost impossible to tell where the dark, grey landscape of the Burren ends and the sky begins.

The name derives from the Irish word *boireann*, meaning 'large rock' or 'rocky district'. The English politician and military leader General Edmund Ludlow, while passing through on a military campaign in 1659, described the Burren as 'a country where there is not enough water to drown a man, wood enough to hang one, nor earth enough to bury him'. In the following centuries, the Burren has been described variously as a moonscape and

Opposite top left: Spring gentian.
Opposite top right: Bee orchid.
Opposite bottom left: Mountain avens.
Opposite bottom right: Yellow wort.

Above: Bloody crane's bill and common spotted orchid.
Right: Dense-flowered orchid.
Below: Irish saxifrage.
Below right: Wood avens.

a fertile rock; songwriter Luka Bloom sang about the 'flowering desert', and for botanist Charles Nelson, the Burren was a 'limestone wilderness'. Whatever you would like to call it, the Burren is a landscape that captures the imagination; it is cherished and respected by its people and known to leave a deep impact on its visitors. The Burren seems to exist on another plane, not quite part of the real world, timeless in one way but also laden with the burden of history. There is just no other place like it.

* * *

The Burren is a limestone karst landscape. Its foundations were laid during the Lower Carboniferous period some 350 million years ago, far away from its current location on the west coast of Ireland. Imagine a shallow ocean in a tropical climate just south of the equator. Some of the inhabitants of this ocean were the ancestors of our snails, mussels and corals, all of which shared one feature: a shell made of the minerals calcite and aragonite. Once the inhabitants of these shells met their end, the shells sank to the seafloor. There, they accumulated, layer after layer, and over a period of some twenty million years, they hardened into limestone, which eventually reached a thickness of up to eight hundred metres. Periodic outwashes of mud, sand and clay from adjacent river estuaries added bands of shale and other deposits that in places separated the limestone beds. During the Upper Carboniferous period, these fluvial outwashes became more dominant, and as a result shale and sandstone built up over the limestone and, in time, completely covered the older rock.

Time passed, continents moved, and Ireland took its current place on the map. The earth's climate oscillated, cooling and warming in cycles. The colder periods saw vast glaciers advancing from the poles, covering much of Europe, including Ireland and the area of the future Burren. Once the temperatures started to rise, the ice would retreat northward for a while, releasing the land from its icy grip. This process went on for thousands of years, and the back and forth and back again movement of these immense masses of ice, several hundred metres thick and weighing many tons, sculpted the landscape. In some places the glaciers cut deep valleys in the rock; in the Burren, they completely removed the top layers of shale and sandstone,

The Burren seems to exist on another plane, timeless in one way but also laden with history

exposing some 250 square kilometres of the underlaying limestone. The best place to see the transition from limestone to shale and sandstone is to the south-east of Doolin Harbour. Here, the limestone disappears under the sandstone cliffs that rise further south to form the Cliffs of Moher.

The glaciers carried vast amounts of rock, gravel, sand and silt with them. Once the temperatures rose and the ice began to melt, this luggage was dropped on the landscape. Drumlins – small, elongated hills of silt, gravel and rocks – are a common sight all over Ireland. In the Burren, drumlins mainly appear along the southern and eastern borders. Erratics, however – from the Latin *errare*, 'to wander' – are obvious all over the Burren, from the coast to the mountaintops. Most of these wandering rocks are made of limestone, but sandstone and granite erratics, which originated further north in Connemara and were transported by the glaciers to their current resting place, can also be found.

The bleak landscape of the Burren that we see today is not an entirely natural phenomenon. After the end of the last glaciation, about ten thousand years ago, the Burren didn't look very much different to the rest of Ireland at the time. After the ice had disappeared, a tundra-like landscape developed; with a warming climate, shrubs and trees established themselves, and over time, vast pine and oak forests took over the land.

It is possible that the limestone here never had a very deep soil cover and was more susceptible to erosion than other

The Burren has one of the highest concentrations of ancient monuments in Ireland, many dating back to the stone age

areas, but the first Burren farmers and the following generations certainly sped the erosion process along by clearing land for crops and livestock. The early settlers also left their mark on the landscape in other ways. The Burren has one of the highest concentrations of ancient monuments in Ireland, many dating back to the stone age. Cooking places known as *fulachtaí fiadh*, portal and wedge tombs, cairns, ring barrows and stone forts, early Christian monasteries, medieval tower-houses and the ubiquitous dry-stone walls are by no means natural but have all become an integral part of the Burren landscape.

Another factor that had significant influence on the juvenile Burren, even before the arrival of man, was water. Rainwater contains liquified carbon dioxide, which makes it slightly acidic; this acidic solution converts the calcite in the limestone into calcium bicarbonate, which dissolves in the water and is washed away. It is this process that shaped and sculpted the Burren to its unique appearance: smooth limestone blocks, or clints, that are separated by deep fissures known as scalps (from the Irish word *scailp*, meaning

Above left: Hazel woodland in summer.
Above right: Flooded hazel woodland in winter.

'fissure' or 'cleft'), as well as grykes, runnels and flutes, terraced hills, sinkholes and vast cave systems. To date, some sixty kilometres of cave passages have been explored, but the true extent of the Burren cave system is believed to be a multitude of that.

The exploration of this underworld brought some unexpected glimpses into the Burren's past. At Aillwee Cave, one of the two show caves of the Burren, the skull of a brown bear was found and dated to be 10,400 years old. A bear patella, discovered in another cave, shows parallel cut marks that were likely caused by a stone knife some 12,500 years ago. The presence of a top predator like the brown bear and signs that it was hunted by men suggest that a diversified flora and fauna already existed in the Burren towards the end of the last glaciation, before the complete disappearance of the ice. The second show cave of the Burren, Poll-an-Ionain, today better known as the Doolin Cave, features the Great Stalactite. With a length of 7.3 metres, this is the biggest free-hanging stalactite in Europe.

Most of the Burren's cave systems have been formed by streams and rivers. The majority of these water courses stay underground; the only stream that runs its complete course on the surface is the Caher River, which has its source beneath Slieve Elva and enters the sea at the

Above: Early autumn in the Burren uplands near Aillwee.
Opposite top: Lesser horseshoe bats in a Burren cave.
Opposite bottom: Cave reflections, Marble Arch Caves, County Fermanagh.

Fanore beach after a rather short run of only seven kilometres. The upper part of the Caher is a gentle flow through hazel scrub and meadows, one of the most secluded and beautiful spots in the Burren. In the early 1990s, local man John MacNamara turned this area into a nature reserve. His goal was to let nature run its course and leave the plants and animals in this little kingdom undisturbed. I had the pleasure to know and work with John for a brief time, but unfortunately our plans to put together a photographic record of the plants and animals at the Caher Valley Nature Reserve were cut short. John passed away after a short illness in 2004. His hopes for the reserve, however, came true: after his death, the gates to the valley were closed, and the plants and animals were left in peace.

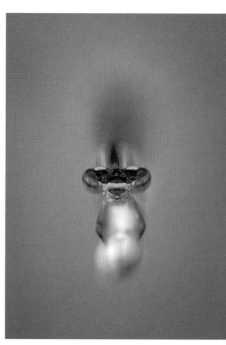

Top left: Lough Bunny in the eastern Burren.

Top right: Common blue damselfly.

Bottom left: Turlough dandelion at Lough Bunny.

Bottom right: Whooper swans, winter visitors, on a turlough in the eastern Burren.

After leaving the Caher Valley Nature Reserve, the river turns west and starts to descend. Trout can often be seen swimming in the clear water, and it is not unusual to come across a dipper, otter or grey heron. On its last two kilometres before reaching the Atlantic Ocean, the Caher drops sixty metres and completely changes its attitude. The gentle river becomes a mountain stream that after heavy rain turns into a raging torrent, rushing around boulders and tumbling over cascades. Why the Caher River never wanders underground is not entirely clear, but it is known that in parts of the Burren, impervious layers of clay, chert and sandstone exist between the limestone. The most likely explanation, therefore, is that the Caher River runs above one or more of these layers, which prevents the water from going underground.

Some rivers make a sporadic appearance on the surface. The Rathborney River emerges from under Gleninagh Mountain, runs down the Feenagh Valley and disappears under the limestone once it has reached Ballyvaughan Valley. Castletown River appears in the Carron Basin, where it is the main source for the Carron Turlough, the largest of the Burren turloughs.

Turloughs are special water bodies that are unique to the Burren and other karst areas. They are often described as seasonal lakes since their existence is very much dependent on the prevalent precipitation. Unlike proper lakes, turloughs are not fed by a constant water supply like a river or groundwater. They are, however, connected to the cave systems of the Burren. During times of consistent or high rainfall, these underground passages flood and eventually overflow, filling the turloughs in the process. During dryer periods, mostly in spring and early summer, water levels fall and the turlough disappears.

All standing water bodies are restricted to the eastern Burren, an area known as the Burren wetlands. In addition to the turloughs, there are fens and lakes. Some are a combination of lake and turlough, so they never disappear completely but will easily triple in size during times of high rainfall. The most interesting of those is Lough Gealain. This lake is fed through springs under the limestone but also features a number of swallow holes that connect to the underground cave system. In autumn and winter, when rainfall amounts are usually higher, Lough Gealain expands considerably and at times floods the nearby road.

* * *

Lough Gealain sits in the heart of the Burren National Park, an area that hosts all the major Burren habitats. In addition to the fens, lakes and turloughs, there are vast stretches of limestone pavement interspersed with patches of fertile soil, small woodlands and wildflower meadows. It is in these habitats where the famed Burren flora can be found. This wonderworld of wildflowers is rather unexpected in a place that on first sight appears to be made of barren and desolate rock. The Burren, however, hosts six hundred species of wildflower and other plants — seventy per cent of all species known in Ireland.

The most famous of the Burren flowers — and something of an emblem for the area — is the spring gentian. With its deep-blue flowers, this is an alpine species which usually flourishes in the high mountains. In the Burren, it bursts into bloom around April and can be found not only at sea level but even right beside the sea. The coastal area of Ballyryan, north of Doolin, is, together with the Burren National Park, one of the strongholds of this enigmatic plant.

Each plant produces several hundred seeds, which are shed in summer for germination the following spring. There is, however, one problem. In order to germinate, spring gentian seeds have to experience frost. Although sub-zero temperatures are not unknown in the Burren, they are very rare, and years can go by between one frost event and the next. In order to survive, the Burren gentian came up with another solution: mature plants produce stolons (underground extensions of the stem), which grow new rosettes and subsequently produce new gentian flowers. After eight thousand years of isolation, the Burren gentian has not only developed this alternative reproduction method, it is also

The Burren hosts six hundred species of wildflower and other plants — seventy per cent of all species known in Ireland

visually slightly different from the populations on the continent. For now, these differences are minute and only of interest to the keen botanist. But give it another few thousand years, and *Gentiana verna* might develop into a completely new species.

Often found in the vicinity of this alpine beauty is the early purple orchid, a species more commonly at home in the Mediterranean. A little bit later in the year, the mountain avens, an Arctic species, joins the mix and bursts into bloom beside the fragrant orchid, the pyramidal orchid, the red helleborine and the twayblade, to name but a few of the twenty-four species of orchid that call the Burren their home.

Another piece in the floral puzzle of the Burren is the presence of lime-loving (calcicoles) and lime-hating (calcifuges) plants, which are often growing in close proximity. Given the ubiquitous presence of limestone, it would be logical to conclude that the soil cover would consist of weathered limestone rock and the remains of glacial drift, a soil type known as rendzina. Calcifuges, however, wouldn't thrive on such a soil. The existence of other soil types in the Burren was a bit of a mystery until microscopic examinations revealed the prevalence of an aeolian (wind deposited) soil known as loess. This loess most likely came

Above left: Lough Gealain and Mullach Mór after heavy rain in late summer.
Above right: Lesser spearwort at the dried-out Lough Gealain turlough in summer.

from east Galway and Connemara and was deposited in various places all over the Burren during the tundra period, shortly after the last glaciation. Another theory is that sandstone and granite erratics that originated further north disintegrated over time to form loess.

Among the other floral oddities to be found here are O'Kelly's spotted orchid, a white subspecies of the common spotted, and the pyramidal bugle, something of a stumped version of the common bugle. The latter is one of Ireland's rarest plants and only flourishes in a few spots along the Burren coast, the Aran Islands (which are geologically a part of the Burren), in Connemara and reportedly on Rathlin Island, off the coast of County Antrim.

The question of why the Burren hosts these unusual and rare plants and plant communities has never really been answered. It is thought that the alpine and Arctic species arrived during or shortly after the last glaciation and were a feature of the early tundra-like landscape. After the climate had warmed enough, other plants, including the beloved orchids, followed from the south and settled beside the existing flora in the mild climate of Ireland's west coast.

* * *

The connection between human activity and the Burren flora became clear in the final decades of the last century. Back then, in order to protect the delicate wildflowers, it was decided to restrict farming activities in the area and considerably reduce the number of grazing animals. Surprisingly, this conservation effort had the exact opposite effect;

because fast-growing grasses and shrubs, mainly hazel, blackthorn and hawthorn, were no longer kept in check by grazers, many of the delicate wildflowers were pushed out of their habitat.

Then, in the year 2010, the Burren Life project was founded, which aimed not only to find a balance between farming and conservation but to actually use farming as a conservation tool. The farmers who signed up for this project pledged to preserve the natural as well as the built heritage of the Burren by re-building dry-stone walls, removing shrub and reintroducing traditional farming methods like winterage, where livestock are brought to upland pastures for the winter months. Today, the Burren Life project is a vital part of the conservation of the whole area. Without grazing animals, the diversity of the flora would be greatly diminished; in turn, the insect population would suffer, which would subsequently impact birds and mammals.

The Burren is a haven for insects. In addition to bees and bumblebees, dragonflies and damselflies, there are twenty-seven species of resident butterfly and well over two hundred species of moth. One of the latter, and a true Burren speciality, is the Burren Green. This moth was first recorded in 1949 and put the Burren on the map for lepidoptera fans; it is the only place in Ireland or Great Britain where this emerald-green beauty has ever been seen.

One of the animals for which insects are a major food source is the bat. The Burren is an important stronghold for these flying mammals — all seven Irish bat species have been reported here, including the lesser horseshoe bat, which is already extinct in many parts of Europe.

Other insectivores are the viviparous lizard, which can often be found sunbathing on the warm limestone; the slow worm, which was introduced to the Burren in the 1970s; and the common frog, who looks slightly out of place hopping across the limestone pavement. The pine marten, whose varied diet includes birds, small mammals, insects, fruit and nuts, and its relative the stoat are the Burren's top predators, while the feral goat is the largest of the wild animals. These descendants of once domesticated goats are not only pretty to look at, with their rugged and colourful appearance, but are also the only animal that munches on the ever-spreading shrubs, thereby playing their part in the conservation of the area.

Unlike other landscapes and habitats in Ireland, the future of the Burren karst and its famous flora is safe, for the moment at least. The combined efforts of farmers, communities and environmental groups make sure of this. The special case of the Burren, however, raises the more fundamental question of what conservation

Without grazing animals, the diversity of the flora would be greatly diminished

Above: Winterage.

really means. Is it right to protect an admittedly unique but ultimately man-made piece of landscape — or would it be proper to let nature reign, which would transform the Burren into a very different landscape? Ungrazed, the Burren would quickly develop a hazel and ash woodland, a process that can already be observed at the foot of Mullach Mór and around Eagle's Rock. In time, other trees, most likely oak and scots pine, would take over, and soil would build up and revert the Burren to its origins as a mixed woodland landscape. Parts of the current Burren flora would survive in this new habitat; others, first and foremost the Arctic and alpine species, would probably disappear.

Ultimately, the Burren as it is today will only exist for a glimpse in the vastness of time. In another few thousand years, the limestone will be gone, washed away into the ocean, or the Burren will have been decimated to a few small limestone islands by rising sea levels. Whichever way it may turn out, another of nature's circles will have been completed.

Above: Common spotted orchid in a hay meadow.
Opposite: Wild garlic in a hazel forest.

Chapter 2

Between the Tides

Imagine the view from a cliff top in the west of Ireland. The vast expanse of the Atlantic Ocean stretches to the horizon, where it merges with a misty sky. Gentle waves travel over this waterscape and disperse in millions of white droplets when they hit the rocky shoreline. The water doesn't mind this measuring of forces as it is always the rock that loses out, one grain at a time. The water finds the weak spots in the rock and over time carves out tunnels, caves and pools.

At the foot of the cliffs lie flat sheets of rock that gently slope into the ocean. Parts of these rock platforms are covered in seaweeds, which shimmer in various green, brown and red tones. Where the seaweed couldn't take hold, ivory-coloured barnacles and black mussels are standing closely packed, giving the rock surface a chessboard-like appearance. Hollows in the

Opposite: Rocky shore, County Clare.
Below left: Thrift on a rocky shore.
Below right: Sea campion on a shingle beach.

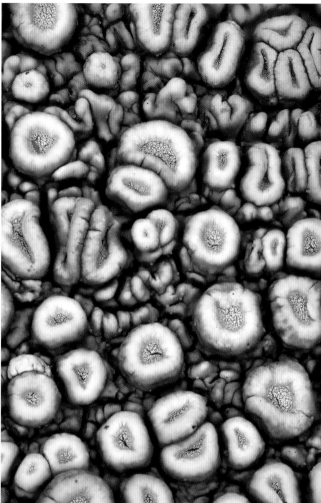

Above left: Lichen species in the splash zone.
Above right: Crab-eye lichen.

rock, some shallow and others knee-deep, are filled with seawater that the retreating tide has left behind; on close inspection, it reveals myriad shapes and colours.

This is the intertidal zone, a strange world teeming with life that belongs fully neither to the sea nor the land. It is a constantly changing and rough environment, full of challenges for its inhabitants. Over the course of a day, they are beaten, suffocated, desiccated, baked and boiled, frozen, hunted and nibbled on.

The force of the pounding waves and the coming and going of the tides are the main problems these rocky shore dwellers have to deal with, and they have developed tools and behaviours to cope with these threats. Body shape helps; being round or flat reduces the area of the waves' impact and leads the water around the body instead of offering resistance. As does having a strong foothold, whether by using suction or by secreting a glue-like substance – some species do both.

To avoid dehydration, those rocky shore inhabitants capable of moving around hide under seaweed, in rock crevices and rock pools – anywhere that some moisture and shade remain after the tide has gone out – and there they wait for the water to return. Going into hiding also helps to avoid being picked up by a hungry predator. Another solution to the dehydration (and predator) problem is having a shell. A shell protects from evaporation, and the space between the body and the shell, known as the mantle cavity, can be used to store water which keeps the body moist and is also a source of oxygen.

Living conditions and the resulting problems differ greatly throughout the intertidal area, and most plants and animals have adapted to a specific part or zone of the habitat. The rocky shore can be divided into either biological zones or physical zones.

Biological zones take their name from the prevailing colour the rock has been clad in. The orange zone hosts lichen of the orange and yellow kind as well as salt-tolerating plants; the black zone is dominated by black tar lichen; the grey zone can be identified by its barnacle and limpet population; and the brown zone is home to the various brown seaweed species.

Physical zones, meanwhile, are defined by their immersion or emersion time. They are measured in metres above CD (chart datum), which is the waterline at the lowest spring tide. The sublittoral zone sits at the bottom of the shore and is always covered with water, even at the lowest spring tide. The lower shore only gets uncovered at spring tides, the middle shore is always exposed at low tide and covered at high tide, and the upper shore is immersed only at spring tides. The splash zone sits above the highest spring tide water mark, and while plants and animals in this zone might get splashed with seawater, this zone never gets covered in water.

* * *

The splash zone is a good place to sit down and have a look over the rocky shore before exploring it in more detail. In spring, this spot is likely to display one of Ireland's most delightful wildflowers. Thrift, also known as sea pink, is a common coastal plant that covers roadsides, cliff tops and dune edges in various shades of pink. Some flowers go to extremes and appear completely white or an almost crimson red. Single plants can grow out of narrow crevices or shallow hollows in the rock, and they are not an unusual sight on walls. Often close by, but far less conspicuous, stands rock samphire, the leaves of which have traditionally been pickled or used fresh in salads; and scurvy grass, a perennial with tiny white flowers and fleshy leaves. These leaves are high in vitamin C, and before citrus fruit was widely available, scurvy grass was a staple on ships to prevent the disease that gave the plant its name.

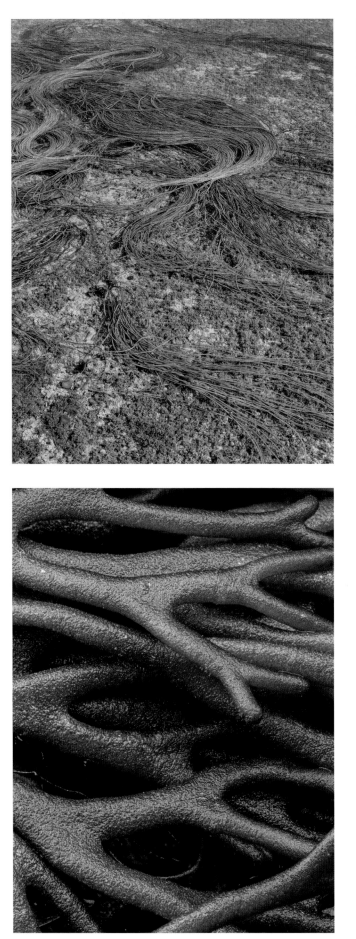

Opposite top left: Thongweed in a rock pool.
Opposite top right: Bright-green sea lettuce and other seaweeds on the lower shore.
Opposite bottom left: Velvet horn.
Opposite bottom right: 'Salad bowl' of seaweeds on the lower shore.

The rocks surrounding these flowers are often covered in a colourful and intriguing mantle of lichen. Around for some four hundred million years, in the past lichens were classified as plants, but today they are recognised as a life form — or rather a complex community of life forms — in their own right. For a long time, it was thought that a lichen was a simple symbiosis between one fungus and an alga, where the fungus provides housing and water in return for food in the form of sugars produced by the alga through photosynthesis. Only in recent years did it become clear that at least some lichen species consist of more than just one fungus and one alga; some have a second fungal partner, which is often a yeast, that is thought to be responsible for the structure of the lichen. Many lichens also have cyanobacteria or other bacteria as part of their community for functions like nutrient transfer between the symbionts or defence against outside threats. Lichen grows very slowly but can live a long time — one hundred years or even older is no rarity — and survive in extreme environments. A sample that was brought to the International Space Station survived for fifteen days in the vacuum of space, enduring regular temperature swings from -12 to +40 degrees.

Seaweeds inhabit the complete intertidal zone as well as the shallow sublittoral zone and are part of the algae family, which, in the widest sense, is part of the plant world. They come in a variety of shapes and colours, from bright green to dark brown, and are divided into three major groups: red, brown and green seaweeds. Red seaweeds are found mainly on the lower shore and can be traced back some 1.2 billion years, making them the oldest seaweed group. Their red colour comes from the special pigments that allow them to photosynthesise at the low light levels that regularly occur on the lower shore. Two common species here are Irish moss, also known as carragheen, and pepper dulse.

The brown seaweeds are the big ones, the likes of the wracks and the kelps that mainly thrive in the colder and nutrient-rich waters of the north Atlantic. They are the youngest of the seaweeds, having developed some two hundred million years ago, and contain a pigment and chlorophyll C to give them their brownish appearance.

Green seaweeds, meanwhile, have the pigment chlorophyll B. Unsurprisingly,

Seaweeds inhabit the complete intertidal zone and come in a variety of shapes and colours

given their colour, they are very closely related to land plants, and many species thrive in freshwater far away from the coast. The most common of the few marine species are sea lettuce and gut weed, which are both the main food source of the rocky shore grazers.

To stay in place, most seaweeds anchor themselves with a holdfast, which is a structure resembling a small plate, often with short, root-like fingers. They also secrete an adhesive compound made of polysaccharides and proteins. As seaweeds are the only plants around, they are an obvious food choice for the grazing animals on the shore. To avoid being constantly nibbled on, some produce anti-grazing compounds like tannin and terpenes, which make them unpalatable. But not only that, seaweeds can also warn their neighbours of potential attacks by periwinkles and other grazers, in the same way that trees and other land-based plants communicate with each other.

* * *

Once the tide has retreated, you can leave the splash zone and make your way further down the shore. First to catch the eye is the ubiquitous common limpet, a member of the mollusc family. Limpets, like many of their relatives, consist of a head with tentacles, eyes and the radula; a muscular foot; and a visceral mass that houses the animal's organs and is covered by a mantle, which is not only connected to the shell but also grows it. The most intriguing part of the limpet's body is the radula. It is effectively a ribbon-like tongue spiked with countless tiny teeth made of goethite, an iron-based mineral and one of the strongest materials on earth.

Limpets, and the majority of the rocky shore molluscs, are grazers. This doesn't mean they are munching on seaweed; they rather scraping up the biofilm that covers the rock and consists of micro-algae, algae spores and cyanobacteria. The main grazing time for limpets is at night when the tide is out. During the day and at high tide, they stay put in their personal parking space. To get the perfect fit, the limpet either grinds away at the rock until it fits its shell or grows its shell to fit the rock surface. The limpet returns to this particular spot after every grazing walkabout. For a long time, it was thought that limpets have a repetitive grazing regime and follow their own trail to get back home, but studies have shown this is not the case. Limpets have no fixed grazing pattern; they take different routes every day and rarely return home the same way they left. Experiments have also shown that they can find their way back even when the surface around their home base has changed during their

Opposite top left: Common limpets.
Opposite top right: Blue-rayed limpet.
Opposite bottom: Acorn barnacles, common mussels and common limpets.

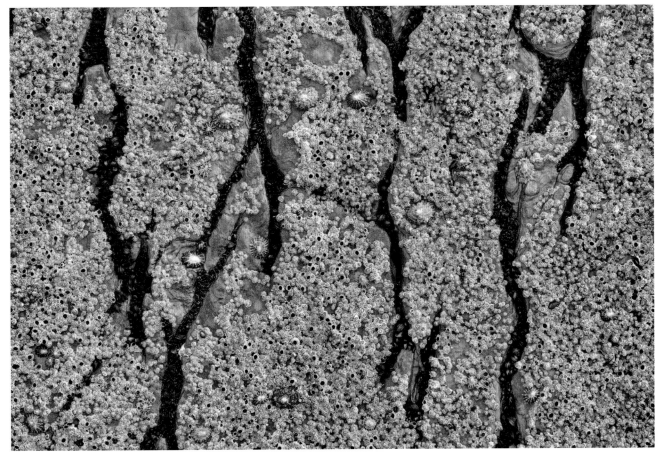

absence, or when obstacles are put in their way. This speaks to a certain amount of topographical awareness, perhaps surprising considering limpets don't have a conventional brain.

What limpets are well known for is their ability to cling onto the rocky surface of their habitat. This is done with the beforementioned muscular foot and also a very special mucus compound that can be switched from lubricant (when on the move) to superglue (when in danger from predators) in seconds. When clinging on isn't enough, limpets have another, rather unexpected, defence mechanism that works especially well against starfish. This technique is known as mushrooming: the limpet lifts its shell up and at the right moment — when the starfish is trying to get to the limpet's exposed body — the shell is brought down hard on the starfish's arm. This hurts the predator, and in most cases, scares it away.

The biggest group of shelled molluscs on the rocky shore are the topshells and periwinkles. Three of Ireland's four topshell species — toothed, purple and grey topshells — have a similar appearance, and it takes a close look to tell them apart. The painted topshell, however, is very easy to recognise, not only because of its beautiful cone shape and colouring that consists of a creamy white with crimson streaks, but also because it looks rather clean compared to other shells. Indeed, the painted topshell has a habit known as shell-wiping. The animal regularly extends its foot all over the shell and wipes it down immaculately. This not only has the obvious cleaning effect, but the painted topshell also gains twenty per cent of its daily food requirement from this action.

The biggest of the periwinkles is the edible periwinkle. As the name suggests, it has been used as a food source for a long time, and even today, bags of boiled periwinkles are a popular snack at many seaside resorts. Like other grazers, the edible periwinkle feeds on the biofilm that covers the rock. However, it adds a little twist to its eating habits: it excretes its food in the shape of pellets and lets bacteria work on it for a while before making another meal of it.

Most of the periwinkles — edible, small and flat periwinkles, the latter coming in a range of colours like yellow, orange, green and brown — live on the middle shore. The rough periwinkle, though, has developed a trick that allows it to live on the upper shore and even venture into the splash zone. It has very small gills, so small in fact that it wouldn't be able to survive under water for too long. Instead, it uses its shell cavity as a lung. Air is stored in the

Above: Common limpet.
Left: Flat topshell.

space between body and shell, and from there, oxygen diffuses directly into the tissue of the animal. The advantage for the rough periwinkle is clear: it has the upper shore and splash zone all to itself, with no other grazers around to share its food.

The easy-to-identify dogwhelk is the only carnivorous shell. Top of its menu are acorn barnacles and the common mussel, but the dogwhelk doesn't shy away from eating other shells – even other dogwhelks. Its eating process is rather fascinating, as well as time consuming. It has a so-called boring organ, which secretes an enzyme that softens the prey's shell. Once enough shell has been softened, the hunter uses its radula to remove the shell fragments, and so on. Once through the shell, it first injects a narcotic to paralyse its prey and then digestive enzymes which will transform the victim into a slurpy soup. The whole feeding process takes time – a day for a barnacle and around a week for a mussel.

One of the dogwhelk's favourite foods, the common mussel, is a member of the bivalve family; that is, molluscs with a two-parted, hinged shell. Bivalves are mostly stationary filter feeders, who anchor themselves into place with the help of a byssus, or strong filament bundle. The common mussel also uses this byssus to bind attacking dogwhelks, which are, if caught, unable to move and doomed to starve to death.

Acorn barnacles, which are often close neighbours of the common mussel, don't have any such defence against predators. These shell-like animals are crustaceans, as are lobsters and crabs, and have evolved to lead a stationary life. Before they settle, however, they spend some time floating around. In their larval stage, they are part of the countless living beings that roam the sea as plankton, microscopic plants and animals that form the base of the oceanic food pyramid. It is not surprising that very few of the young barnacles make it to adulthood – most are eaten straight away or swept out into the open ocean to be eaten there. The ones that make it go looking for a suitable dwelling place,

The easy-to-identify dogwhelk is the only carnivorous shell

and they are rather picky. The rock or other underground that will become their permanent home has to be smooth, not covered in too much biofilm, and it must be in a location that provides floating food. Once a place has been selected, the barnacle attaches its head to the underground and starts building its cone-shaped house, which consists of six overlapping calcareous plates that leave a diamond shape opening on the top. When the tide is out, this opening is sealed by two more calcareous plates; when the tide is in, the barnacle extends its feet out of the opening to catch any floating, edible debris.

* * *

While most shore dwellers prefer to blend in and therefore display a rather understated appearance, there are other animals that immediately stand out. Sea anemones fit this category, and the most common among them is the beadlet anemone. This beautiful animal comes in a variety of colours like red, green and brown. The similar strawberry anemone is a different species, and there is increasing evidence that the brown and green beadlets might be as well. Despite its pretty, flower-like looks, the beadlet anemone is a ferocious hunter and very territorial. Should another anemone come too close, a headbutting fight breaks out, each trying to get close enough to deliver painful blows with its blue stinging cells, known as nematocysts, which act as miniature poison-filled harpoons. These battles can take many days until one anemone finally has enough and retreats. The main use of the stinging cells is, however, to stun prey, which can be a crab, prawn or jellyfish. The stunned victim is then directed by the tentacles to the beadlet anemone's mouth, which sits in the centre of the oral disc. Once the prey has been digested, its remains take the same way out as they came in. Most beadlet anemones reside in rock pools, but they can survive outside water for a certain time by retracting their tentacles and closing their mouth and oral disc to reduce the evaporation rate. Beadlet anemones live up to three years in the wild, but in captivity they can get much older. Granny, a beadlet anemone that spent part of its life in an aquarium at the Royal Botanic Garden Edinburgh, became a bit of a celebrity by reaching the age of fifty.

While the beadlet anemone is a bit of a loner, the snakelocks anemone prefers company and lives in a colony. One reason for this is its preferred reproduction method: it simply splits in two, a process known as longitudinal fission. This means any snakelocks anemones that sit very close together are most likely clones. The snakelocks come in two varieties, a brown-coloured one and a green version with purple tentacle tips. The latter hosts a symbiotic alga that supplies additional food to its host, and in return the anemone provides shelter and certain nutrients to the alga. Naturally, this version of the snakelocks anemone prefers shallow waters and sunlit pools in which the alga can photosynthesise. To avoid any damage from sun exposure, the anemone produces fluorescent proteins that act as a sunblock and are responsible for the purple colouring.

One of the most beautiful members of the anemone family is the dahlia anemone, which can on occasion be found in the bigger and deeper pools. This colourful anemone can reach a diameter of up to 15cm and form large colonies that transform the rock into a deadly flower garden.

Urchins are a common sight in rock pools. These animals belong to the echinoderms (meaning 'spiny skinned'), an animal group that can be traced back some five hundred

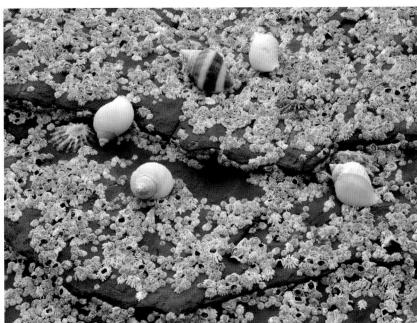

Opposite: Painted topshell.

Top left: Edible periwinkles.

Top right: Periwinkle eggs.

Above left: A dogwhelk attack survivor, with the borehole visible on the shell.

Above right: Dogwhelks and young (white) and older (yellow) acorn barnacles.

Opposite: A beadlet anemone with its prey, a common limpet.

million years. The urchin's body is shaped from fused calcareous plates that contain the internal organs and are covered with numerous sockets that hold protective spines. The outside of this shell, known as the test, is enveloped in a thin skin, pedicellariae (mini-pincers used to keep the animal tidy) and tube feet, which are responsible for locomotion, feeding and respiration. Urchins are grazers and have a mouth, known as Aristotle's lantern, that is a very effective contraption made up of plates, muscles and chisel-like teeth to scrape the rock clean of any edible material.

Black sea urchins, also known as purple sea urchins, often form colonies in shallow rock pools and can sometimes be seen with limpet shells or pieces of seaweed on them. This is no accident: the urchins are using them as sun protection. Black sea urchins are also known to scrape out shallow burrows from the rock, which they use as a resting place – they're also known as the 'rock-boring' urchin. The black sea urchin is a prime example of what overfishing can do. In the second half of the 20th Century, black sea urchins were harvested and exported from Ireland on a large scale. In 1976 alone, 350 tons were landed; ten years later, this was already reduced to forty-eight tons; and another ten years on, only six tons could be harvested. A few years after that, the black sea urchin had virtually disappeared from the Irish coast. Since then, the animal has made a slow recovery and is recolonising the west coast, but its abundance is still a far cry from the numbers that could be seen on Irish shores fifty years ago.

An occasional visitor to the intertidal zone is the edible sea urchin. This urchin is considerably bigger than the black sea urchin, has shorter spines and comes in a colour range from white to red. This species lives in the deeper waters of the sublittoral zone, and its visits to the middle and upper shore are rather unintentional, but from time to time an animal gets trapped in a rock pool and has to wait for the next high tide.

Starfish are also members of the echinoderm family. Their plates are less rigid than those of sea urchins and are held together by connective tissue, which gives the starfish greater flexibility. Just like edible sea urchins, they are not regular inhabitants of the intertidal zone but sometimes get stranded in rock pools or under rocks. The most common starfish along the Irish coast are the seven-armed starfish, obviously named for its unusual number of arms (most starfish have only five); the spiny starfish (a big, greenish starfish that lives up to its name); the bloody henry starfish, which is usually a purply-red; and the common starfish. The latter feeds mainly on mussels and is the

The black sea urchin is a prime example of what overfishing can do

most likely to appear outside the sublittoral zone. Once it has targeted its prey, it uses its tube feet to force the two halves of the mussel apart, and when it has a good grip, it inserts its stomach right into the mussel. This appetite got the common starfish in trouble with fishermen; an old fable tells of fishermen ripping the starfish in half to protect their shellfish stock, only to see themselves confronted with even more starfish a few days later. The reason for this is the unusual regeneration ability of all starfish – one arm and part of the central body are enough to regrow a complete animal.

* * *

These creatures make up only a tiny percentage of all the life that thrives on rocky shores. Many rocky shore inhabitants are tiny, barely visible to the naked eye, like shrimps, sea spiders, midge larvae, marine springtails and minute brittle stars. Others are masters of disguise, like numerous fish species, worms and crabs, including the hermit crab, a shy but curious animal that uses the discarded shells of periwinkles or topshells as its home. Other life forms are hardly recognisable as such: squirts, mats and sponges. The latter are among the most primitive life forms on the shore.

It's a wondrous world, the no man's land between the low-tide and high-tide marks, with many undiscovered secrets. Unfortunately, the communities living here face not only the problems of waves and tides but other threats creeping up on them from all sides. Climate change will bring warmer and more acidic water and more frequent and violent storms. From the land, the intertidal zone faces untreated sewage and run-off from heavily fertilised fields. Plastic chokes the shores; thoughtlessly dumped bottles and wrappers from one side, and the constant supply of flotsam and jetsam that the tide brings in from the other. It is not only the bigger pieces that can entangle animals or be mistaken as food and ingested by birds, fish and other animals, eventually killing them. The greatest threat comes from the barely visible plastic particles, so-called microplastics, some of which are small enough to be devoured by plankton. These accumulate throughout the food chain, slowly releasing their toxic components. The old saying 'what goes around comes around' has never been illustrated more clearly — the future for the rocky shore dwellers, and for ourselves, is uncertain.

Opposite top left: Beadlet anemone.
Opposite top right: Jewel anemones.
Opposite bottom left: Snakelocks anemone.
Above right: Strawberry anemone.
Above: Green variety of the beadlet anemone.

Top left: Chiton.

Top right: Shore crab hiding in sea lettuce.

Above left: Edible sea urchin.

Above right: Rock pool with black sea urchins.

Opposite: Common starfish.

Top: Spiny starfish.

Above: Sea hare.

Opposite top left: Rough periwinkles.

Opposite top right: Breadcrumb sponge.

Opposite bottom left: Compressed purse sponge.

Opposite bottom right: Rocky shore community under a rock overhang.

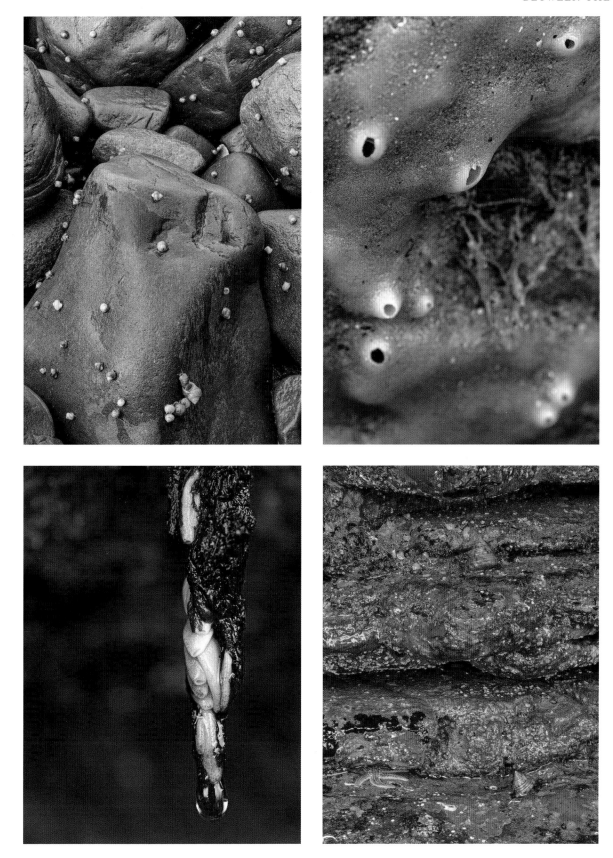

Above: Rocky shore at Clahane, County Clare.
Opposite: Rocky shore at Fodry, County Clare.

Chapter 3

Along the Shannon

The Shannon has been a part of my daily life now for over twenty years. I only have to go down the road to sit at its shores at the small harbour at Kilbaha Bay, where the Shannon Estuary opens up into the mouth of the Shannon. If I travel east, I can witness the estuary narrowing and changing its appearance, transforming from a choppy ocean inlet to a gentle river. There are few experiences more wonderful and calming than spending a summer morning on its banks, be it along the river walk at O'Briensbridge, the meadows near Clonmacnoise, or one of the numerous small harbours like Termonbarry, Roosky or Drumsna. Darkness slowly fades, and the first shimmer of light reveals clouds of mist rising from the slow-moving water. On the opposite shore, the shape of a grey heron becomes clearer as it takes off with a rough croak after being startled by a splash in the water nearby. Maybe an otter diving for

Opposite: Spring dawn at Kilbaha Bay, Shannon Estuary.
Below left: Flag iris at the edge of a Callows field, with an esker in the background.
Below right: Amber snail, Shannon Callows.

its breakfast? The first rays of sunshine bathe the scene in a golden light, illuminating the dew that has formed overnight on the wildflowers, reeds, sedges and grasses. A dragonfly sits motionless on the stem of a yellow-flowered flag iris, waiting for the reviving warmth of the morning light. Birdsong fills the air, and the rising sun is doing its best to burn away the fog that is now rising heavily from the river. In a few hours, the sun will have succeeded, the air will be clear and filled with the humming and buzzing of insects, and the Shannon will glimmer blue in the summer heat.

* * *

Ireland's longest river, the Shannon emerges in the Cuilcagh Mountains at the border between counties Fermanagh and Cavan. From here, the juvenile river makes its way almost

Opposite: A Shannon Callows meadow and hedgerow near Banagher.
Below left: Sedge warbler, Shannon Callows.
Below right: Yellowhammer, Shannon Callows.

Above left: Water forget-me-not, Shannon Callows.
Above middle: Summer snowflake, Shannon Callows.
Above right: Creeping yellow cress, Shannon Callows.

straight south, takes on numerous tributaries and grows quickly while winding its way through the flat landscape of the Irish midlands. Along the way, the Shannon forms three major lakes, Lough Allen, Lough Ree and Lough Derg, and some smaller ones, Lough Cory, Lough Tap, Lough Boderg, Lough Bofin and Lough Forbes, before turning west after passing through the city of Limerick and opening into a long estuary.

Being one of Ireland's major waterways, the Shannon is neither pristine nor untouched. Centuries of shipping traffic, hydroelectric power generation schemes and the general pollution and destruction that follows everywhere humans go have left their marks on many parts of the mighty river. Some spots, however, have, at least to some extent, escaped the forces of modern civilisation and exist as they would have some hundreds of years ago.

One of these spots is the Shannon Callows, the floodplains that got their name from the Irish *caladh*, meaning 'river meadow'. This area of wet grassland extends on both sides of the river between the towns of Athlone (on the southern end of Lough Ree) and Portumna (on the northern end of Lough Derg) and owes its existence to the low gradient of the Shannon. On its journey of two hundred kilometres, the Shannon only drops twelve metres, and this overall shallow gradient is taken to the extreme along the Shannon Callows. Between Athlone and Shannonbridge (which lies about halfway between Athlone and Portumna), the Shannon only drops thirty-five centimetres, which creates a gentle and slow-flowing stream, or, as some have described it, a moving lake.

In 1896, this unassuming landscape was described as 'low, flat, boggy and irksome' by a visiting traveller. The Callows are indeed flat and surrounded by peatlands, but they don't stay irksome for long once you start exploring the vast meadows. The Shannon makes its way through the Callows in many twists and turns and is almost level with the surrounding grasslands – steep riverbanks are a rarity. In places, reedbeds seam the water's edge, giving shelter to some of the Callows' numerous inhabitants. Hedgerows, mostly made up of willow and hawthorn, divide the grasslands into fields and on occasion run right to or even along the water's edge.

The Shannon Callows are one of the last remaining floodplains in Europe. River engineering and drainage schemes have tamed the vast majority of Europe's rivers in the last century. In Ireland alone, six floodplains have been rectified to create more high-yield farmland. The Shannon Callows would have fallen to the same fate if the arterial drainage campaign for the Shannon in the late 20th Century hadn't failed again and again due to planning problems and high costs.

Today the Shannon Callows encompass an area of around thirty-five square kilometres and are being farmed in a traditional way that follows the rhythm of nature. The fields and pastures that make up the Shannon Callows are all in private ownership, and each section has its own name: the Inch Callow, Bridge Callow, Foolagh Callow, Tower Callow and many more.

Towards the end of the summer, rainfall amounts grow and the frequency of downpours increases, drenching the countryside. All this water must go somewhere; it trickles into narrow brooks, which join small streams, which find their way into larger rivers, which eventually feed into the Shannon. As a result, the Shannon bursts its banks and spills over into the fields, usually towards the end of October. At first, only the banks disappear under water, and the drainage channels fill to the brim. Then, with more and more low-pressure systems emptying their load over Ireland, water levels continue to rise; by Christmas, all callow lands have vanished under a vast lake. The rising Shannon, however, brings not only water but also large amounts of nutrient-rich sand and silt, which eventually settle on the submerged grasslands. It is these deposits that fertilise the Shannon Callows and make possible what happens in spring.

The waters start to recede in March, and by mid-May the Shannon has returned to its bed and the Callows are usually free of standing water. With temperatures rising, the grasslands burst into life. The grasslands of the Shannon Callows are classified as seminatural and started to form after the ice age. The Shannon, and probably even the larger lakes, already

The Shannon Callows are one of the last remaining floodplains in Europe

Top left: Dark clouds and sunshine over the Callows near Clonmacnoise.
Top right: Irish hare at the Shannon Callows.
Above left: Spiderwebs and greater knapweed on a summer morning, Shannon Callows.
Above right: Marsh marigold, Shannon Callows.
Opposite top: The Callows at Clonmacnoise.
Opposite bottom: Late summer at the Callows near Shannonharbour.

existed back then and were periodically covered in thick ice sheets. After the end of the last glaciation about ten thousand years ago, the Callows became a large, island-studded lake known as the Extended Lough Ree-Derg, surrounded by eskers, high ridges of sand and gravel that were left behind by the glaciers.

Rising temperatures allowed vegetation to develop, and a sudden drop in water levels

around nine thousand years ago transformed the Callows into a mixture of fen, reed swamp and marshy forest. A relatively dry and settled climate over the following two thousand years saw water levels drop further, and a fen woodland developed on both sides of the Shannon, which consisted mainly of oak, alder, elm and birch, with willows and reed banks growing closer to the river's edge. The Callows as we know them today took shape some four thousand years ago, when a change in climate and increased human activity transformed the landscape once again.

The transition to today's Atlantic climate regime triggered seasonal flooding along the Shannon, which was supported by the removal of trees to create farmland. Some of the fen woodland remained into the 17th Century, but by 1800, all the woodland was gone, and

Above: Floods at Clonmacnoise.

Above left: Flooded Callows in winter.

Above right: Lapwings at the Callows on a winter morning.

Bottom left: Snipe on a cold winter morning.

Bottom right: Whooper swans on the flooded Callows.

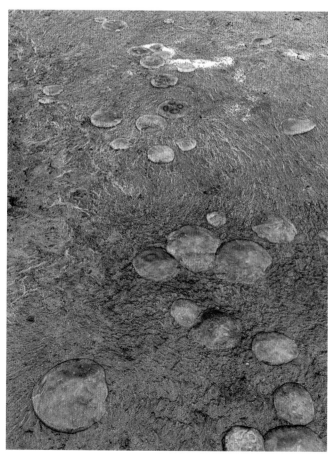

drainage ditches modified the fens and marshes to make them suitable for grazing animals. This created the wildflower meadows of the Shannon Callows, which have never been ploughed or resown. So, although all the plants are native, today's flora of the Callows was shaped by draining, grazing and cropping; it is a man-made habitat, similar to the Burren. Should farming cease along the Shannon, the Callows would quickly revert to fen woodland.

＊＊＊

Over two hundred plants have been recorded in the Callows, including many common ones but also some rarities. While the grasslands look rather uniform at first sight, a closer look reveals different habitats and plant communities. Drains, ditches and the edge of the river support specialist plants that don't mind standing or even floating in water. Common reed is the most obvious and widespread. It's a member of the grass family, which can reach heights of over two metres and was once widely used for thatching roofs. The free-floating frog-bit is less eye-catching but rather intriguing. When an area becomes too crowded or dries out, the plant can produce stolons and travel considerable distances in that manner. After flowering in July and August, the frog-bit dies back, but not before it has produced a 'turion', a bud-like entity that spends the winter on the riverbed to avoid frost. The following spring, the turion grows into a new frog-bit plant. Other water-loving plants include the yellow and white water lilies, the elegant arrowhead, and the rare water parsnip, which only grows along the Shannon and the Erne.

A few steps away from the aquatic habitat starts the wet alluvial grassland. It stays wet and marshy throughout the summer and features a variety of sedges and grasses as well as wildflowers like brooklime (whose leaves are bitter but edible), water forget-me-nots, and marsh marigold (which is also known as mayflower and is supposed to ward off evil when strewn on the doorstep).

Over two hundred plants have been recorded in the Callows, including some rarities

The largest habitat, making up around ninety per cent of the Shannon Callows, is marshy grassland. This type of grassland is somewhat similar to alluvial grassland, but because it spends less time flooded and waterlogged than the land directly adjoining the river, it supports a greater variety of plants. Grasses and sedges are also widespread here, but the variety of wildflowers is far greater: yellow loosestrife, ragged robin, meadowsweet, marsh bedstraw, marsh pea and buttercups are just a few of the species that thrive here. The unmistakable and rare highlight of this type of grassland is the summer snowflake. A tall plant that produces elegant white flowers very early in spring, it is a member of the daffodil family and grows only in a few secluded spots of the Callows.

Dry grasslands extend above the normal flood line and sit on well-draining soils made up of old river alluvium and glacial material. These meadows feature adder's tongue, common spotted orchid, twayblade, cowslip, yellow rattle and other wildflowers that don't like it too wet.

* * *

Unsurprisingly, the rich and varied flora of the Callows attracts much wildlife. Insects and other invertebrates are plentiful, and they feed on and shelter in the rich vegetation. Bees and bumblebees, dragonflies and damselflies, butterflies, snails and others are at home here, but what the Shannon Callows are most renowned for is birds. In spring, the hedgerows echo with birdsong when blackcap, whitethroat, willow warbler, reed bunting, skylark and others establish their territory and look for a mate. The cuckoo can also be heard, and sometimes seen, before it places its eggs into a foreign nest, most often that of the meadow pipit.

The most enchanting soundtrack of the Callows comes from the waders; yelping and yodelling, melancholy ringing and soft chanting floats over the grasslands. The creators of these sounds, the breeding lapwing, redshank and curlew, stay mostly hidden, but one breeding bird that you might get a glimpse of – and the one who produces the most mesmerising and eerie sound – is the snipe. This characteristic sound, known as drumming, is created by the male bird when it dives and the air is forced through two splayed, vibrating tail feathers. Once heard, it is a sound you will never forget. The snipe starts breeding in late April and prefers to build its unique nest, which features a woven canopy, near soft ground where it can forage for food with its long, straight bill.

Redshanks start breeding at the same time and build a simple, grass-lined nest in high grass that keeps the bird out of sight of predators. The redshank's breeding season is usually finished by June, when it leaves for its wintering grounds at the coast or further abroad in Britain and mainland Europe. The curlew breeds in a similar fashion but stays considerably

Opposite top left: Scurvy grass, Shannon Estuary.
Opposite top right: Sea aster, Shannon Estuary.
Opposite bottom left: Yellow horned poppy, Shannon Estuary.
Opposite bottom right: Sea lavender, Shannon Estuary.

Top left: Grey heron, Shannon Estuary.
Top right: Sanderling, Shannon Estuary.
Above left: Great crested grebe, Shannon Estuary.
Above right: Widgeons, Shannon Estuary.

longer, with chicks hatching only in June, which is the reason for its sharp decline in most parts of the country. It is estimated that ninety per cent of all breeding pairs have been lost, and a survey carried out between 2015 and 2017 counted only 138 active breeding pairs. The hay meadows of the Shannon Callows are cut only once and late in summer, around mid-August. This gives all the ground-breeding birds the chance to raise their young. The intensive farming regime which has become common in most parts of Ireland, however, aims for three or even more cuts, with the first one being carried out as early as May. In the best-case scenario, this would leave the birds exposed to predators; in the worst, the breeding parent and its brood could be killed by the heavy machinery.

The fourth of the breeding Callow waders, the lapwing, has a bit of a different breeding strategy. Lapwings breed in groups out in the open, in areas with short grass, where they scrape a simple hollow in the ground. Their chicks arrive around late April, and like all wader chicks, they can fend for themselves almost immediately after hatching but stay under the watchful eye of the parents for another few weeks.

While the Shannon Callows are a haven for these birds, even here their numbers are steadily declining. Still, it is not likely that they will disappear like the corncrake. The Callows were once its stronghold, but for reasons unknown, the characteristic 'crex crex' call that gave the bird its Latin name hasn't been heard here for many years. Usually, the corncrake arrived from its wintering grounds in Africa in April to raise its offspring at the

Below left: Sandwich tern, Shannon Estuary.
Below right: Ringed plover, Shannon Estuary.
Bottom left: Brent geese, Shannon Estuary.
Bottom right: Curlew, Shannon Estuary.

Above left: Querrin Creek and the Shannon Estuary beyond.
Above right: Mudflats at Kildysart, Shannon Estuary.

edge of the hay meadows. The first brood hatched in June, and unlike wader chicks, the young corncrakes were fed by the mother for a few days. A second brood arrived in late July or early August, and by September, the birds would start their return journey to their winter home.

Today, the corncrake is on the edge of extinction, and its Irish breeding population has shrunk to a few pairs that raise their young in remote areas of counties Mayo and Donegal, mostly under the watchful eyes of members of the National Parks & Wildlife Service and Birdwatch Ireland.

In late summer and early autumn, after the summer visitors have vacated the Callows for their wintering grounds, the winter visitors arrive. The Greenland white-fronted goose from Greenland; whooper swan, teal, shoveler, pintail and golden plover from Iceland; dunlin, lapwing and shoveler from Scandinavia; dunlin and Bewick's swan from Siberia; and redshank and greenshank from Britain. Some sixty thousand birds use the Shannon Callows as their winter residence. The flooded landscape provides safety from predators, shelter and food; underground stems of rushes and buttercups, seeds of grasses, and snails and other invertebrates provide lasting sustenance for a long winter.

The flooding that provides the lifeblood for the Callows doesn't always follow a given timetable. In 1983, the Brosna Callows were flooded in June after an unusually wet spring and early summer, while in 1985, the Callows were full by August and the rising waters were surrounding farms and houses by September. Extreme winter flooding also happens on a regular basis, the worst of which occurred in 1955, when water covered four thousand hectares of land, with another four thousand completely waterlogged. Over the past decades, these extreme events seem to happen on a more and more regular basis; above-average flooding has been recorded in 2009, 2015 and 2020. If climate change predictions prove to be right, these extreme floods could become the new normal, which will not only destroy the livelihood of many Callows farmers but also affect the wildlife and change the flora and fauna of this rare place forever.

* * *

Following the Shannon south and past Lough Derg, you will eventually veer westward and reach the city of Limerick. Here, the Shannon becomes tidal, and soon after leaving the city, the river will widen considerably and become an estuary.

River estuaries have always had a special allure for me. These wide expanses that are neither river nor ocean, neither land nor sea, are being shaped by the endless rhythm of the tides on one side and the flow of the river on the other. At high tide, there is just a wide swathe of water; only low tide reveals the intricate pattern of channels and runnels that have been carved into the vast areas of sand and mud by the flowing waters, and this strange landscape is teeming with life.

The Shannon, together with the Fergus River, which joins the Shannon near Ennis, form Ireland's largest estuary complex. The vastness of these mudflats is enthralling. I remember a New Year's morning waiting for the sunrise at Drumquin Point, south of Clarecastle, close to where the Fergus meets the Shannon. Here, the mudflats extend to the horizon. It is a vast world of waterlogged sand and mud, understated but dwarfing everything in its vicinity. Its smooth, reflective surface is only disrupted by the network of channels that grow tree-like through the dense substrate. As light levels rose on this cold and overcast winter morning, the damp surface of the mud started to glow in the colours of dawn: bluish at first, then slowly turning into shades of purple and magenta. Just before sunrise, the expanse glared in rich red and orange tones, then the sun rose over the horizon, quickly disappearing behind the clouds and the estuary returned to muddled shades of brown.

In the dull winter twilight that followed, the sounds of the estuary echoed eerily across the wide-open landscape. The oystercatchers announced themselves with a high-pitched, sharp call, and seconds later a group of birds zipped by, flying close to the surface to find a quiet place for breakfast. The striking colours of the birds, the black and white body and the bright-red beak, were, however, muted under the dark sky. Further out on the mudflats, dark shapes moved here and there in search of food. Some strode gracefully, others scuttled randomly, and yet others stood motionless. The air was filled with chittering, whistling, bubbling and chirping, one of the most beautiful soundscapes I can imagine. The sound of the estuary.

The Shannon Estuary stretches for some one hundred kilometres between Limerick and the Atlantic Ocean, which it meets between Kilcredaun Point on its County Clare side and Kilconley Point in County Kerry. Here, the wide waterway opens even further into the mouth of the Shannon. This expanse of water is enclosed by sheer cliffs, and its entrance is marked by the headlands of Loop Head to the north and Kerry Head to the south. Beyond those headlands lies the open Atlantic.

The Shannon Estuary is a bit of a paradox. For centuries, this waterway has been one of the main routes into Ireland for both passengers and goods, as well as an important fishing ground. While local fishery has been in decline for a while, goods are still being brought into Ireland via the Shannon Estuary and unloaded at the deep-water port at Foynes, which is one of the biggest in Ireland and handles cargo from all over the world. Not too far away from Foynes sits Shannon Airport, one of the major airports of the country, on the banks of the estuary. There is also the controversial Aughinish Alumina plant and two coal, oil and gas fired power stations, one at Moneypoint on the Clare side and one in Tarbert on the Kerry side. Those not only add to the shipping traffic but also pose a constant threat to the fragile habitats of the estuary.

Despite this human presence, the Shannon Estuary is one of Ireland's most intriguing and important sites for wildlife. The sheer cliffs around the mouth of the Shannon are breeding sites for fulmars, kittiwakes, guillemots and razorbills that arrive in their hundreds in early spring, turning the coast into a cacophony of cackling, growling, whistling and twittering. One reason for these birds to choose the mouth of the Shannon as their nesting site is the rich food supply. Herring and sprat are abundant, and these not only attract the local nesting population but also

Rare species like the chough and the raven have established a stronghold here

animals from further afield. Gannets have their nesting site on the Skellig Islands, some one hundred kilometres south, but are a common sight at the lower estuary during the summer.

The shores of the estuary are also home to some permanent residents. Cormorants and shags, along with herring gulls, lesser black-backed gulls and greater black-backed gulls still appear in high numbers, while rarer species like the chough, unmistakable with its red beak and legs, and the raven, with its wonderfully eerie croak, have established a stronghold here.

After the summer visitors have reared their offspring and left the cliffs in a sprinkled black and white pattern — a painting of guano on rock — they return to the open ocean to make space for the winter visitors. The Shannon Estuary features a good number of small islands and sheltered creeks, inlets and bays, most of which reveal wide sandy beaches and vast mudflats once the tide is out. These places are the perfect wintering grounds for waders and waterfowl, providing not only a rich food supply but also protection from the worst of the winter weather. The Brent goose, whooper swan, lapwing, curlew, greenshank, redshank, golden plover, knot, dunlin, wigeon and others join resident birds like the grey heron, little egret, oystercatcher, ringed plover, snipe, shelduck and mallard from October onwards.

From a distance, the mudflat, this expanse of glorious mud and sand, appears dull and lifeless. A closer look, however, reveals an abundance of life. Mudflats consist of very small sand particles, smaller than the grains found on a sandy beach, that have been moulded together by tidal currents. The electrical charge that causes repulsion between the grains is very much reduced in the tiny particles that form the mudflat. This makes them stick together more easily, and the inhabitants do the rest by adding mucus and faeces to the mix. The result is a very stable habitat that can resist the action of the tides much better than a sandy beach.

Most of the mudflat community lives underground. Apart from single-celled organisms like diatoms (a type of algae), cyanobacteria and flagellates, the main inhabitants are worms – including the lugworm and the sand mason – and bivalves – among them the cockle, razor shell and scallop. Both worms and bivalves are an important food source for crustaceans like crabs and lobsters, the various fish species of the estuary, and the waders, who are uniquely adapted to search for this kind of food. Birds with long beaks like the oystercatcher are tactile foragers and actively probe the ground for prey. Birds with shorter beaks like the plover are visual foragers. They wait for signs of activity and only go in when they notice movement

Below left: One of the Shannon Dolphins, Shannon Estuary.
Below right: Grey seal, Shannon Estuary.

under the surface. These different feeding methods become obvious in the behaviour and movement of the birds. While the oystercatcher roams in a rather calm and organised way, probing the ground with every step, plovers are all over the place, running, stopping, looking and running again, without showing any discernible pattern in their movements.

Adjoining the mudflat in many places is a saltmarsh. These areas are often dominated by the common cord grass. This invasive species is a hybrid between the native small cord grass and the smooth cord grass, which was introduced from North America. Initially this hybrid grass, with its dense rooting system and thick growth on the surface, was very welcome and used to stabilise the coastline. Soon, however, the common cord grass took matters into its own hands and spread aggressively all along the Irish coast. Today, it is not only a threat to the native flora but also takes over feeding and roosting sites for waders and wildfowl. Where the common cord grass hasn't yet invaded, the native saltmarsh flora, which includes a number of salt-tolerating wildflowers, still thrives. Throughout spring and summer, a trio of purple flowering plants add a splash of colour to the landscape of the saltmarsh. It starts in April with thrift, also known as sea pink, which often grows side by side with scurvy grass, a small white flowering plant. In summer, sea lavender appears, with its delicate, tiny flowers, followed by the striking sea aster, whose yellow and purple flowers turn into fluffy seed heads towards the end of the season.

* * *

Away from the shore lives a somewhat unexpected group of animals that are locally known as the Shannon Dolphins. They are bottlenose dolphins and form one of only a few resident groups in the whole of Europe. The Shannon Estuary has been their home for hundreds, if not thousands, of years. One of the earliest accounts of the Shannon Dolphins might be found in the local legend of the Cataigh (or Catach), a sea monster that claimed Scattery Island, one of the small islands of the estuary. It was described as an enormous eel with a line of sharp barbs along its back and dagger-like teeth that were long enough to curl around the island. Because there is some truth in every legend, it is likely that there really was a sea monster — or in this case, probably many of them. Witnessing a group of dolphins travelling with their fins cutting through the water, one could indeed imagine an eel-like creature swimming along, and a close look at a dolphin's teeth indeed reveals dagger-like shapes.

Today, over one hundred dolphins are present at the estuary at any given time. Of those, around forty are permanent residents, while the others come and go throughout the year. The Shannon Dolphins travel in groups which can consist of up to twenty individuals but are on average considerably smaller. While these groups are very fluid, known as fission-fusion society, it is generally individuals of the same sex and age that travel together. Groups of mothers with calves, for example, can regularly be seen in the Shannon Estuary.

The reason the estuary is so popular among dolphins is the same as it is for birds: the rich food supply that is being brought in by the strong tidal currents. Bottlenose dolphins in general live on a diet of fish, squid, and sometimes small crustaceans. The Shannon Dolphins are most likely mainly fish eaters, feeding on resident fish like bass and various species of flatfish as well as pelagic species like herring or sprat that venture into the estuary. In late summer and autumn, salmon becomes a major part of their diet, and they can be seen tossing these fish in the air, a behaviour that is probably associated with both playing and learning.

Bottlenose dolphins travel and hunt in small groups and find their way — as well as their prey — through echo location. While a dolphin's eyesight is similar to our own, it is of little use in the often murky waters of the estuary, so echo location is a much more reliable method to get around safely. The dolphins emit a series of pulses and clicks from an area near their blowhole known as the melon. When these sounds hit a solid object like a fish or

Below: Grey heron at Carrigaholt Harbour, with Moneypoint Power Station in the background.

a boat, they bounce back and the dolphin picks up those echoes through its lower jaw, from where they are transmitted to its inner ear. Because sound travels five times faster in water than in air, this way of experiencing the world is very effective. It is also thought that certain variations of those clicks and pulses, along with whistles, are a form of communication between the animals. Tailslapping can also be a kind of communication. Depending on the strength of the slap, this behaviour could be a warning to other animals or an attempt to make contact. Tailslapping has also been observed as part of hunting techniques to stun the prey. Another popular hunting technique that can be observed on the Shannon is a group of dolphins circling a school of fish to keep them together, while individuals repeatedly dash into the tightly packed ball of fish to feed.

Dolphins are not the only marine mammals in the Shannon Estuary. Other cetaceans, especially minke whales, regularly find their way into the area, and bigger whales like humpbacks and orcas can be seen passing or feeding off the headland of Loop Head. Common seals are regular visitors, while grey seals are known to rear their young in the small storm beaches at the mouth of the Shannon and the lower estuary. Unlike common seal pups, who can swim within hours of birth, the offspring of the grey seal is born with a white fluffy coat that needs to be shed before the youngsters can enter the water. This takes up to six weeks, and during this time both mother and pup are stationary, the pup hauled up on the beach while the mother patrols the waters around the cove. Seals are the source of another legendary figure, the selkie, which appears not only in local tales but legends all over Ireland and Scotland. Usually the selkie,

Humpback whales and orcas can be seen passing or feeding off the headland of Loop Head

like a mermaid, is trying to seduce and then cause harm to a fisherman that has done her wrong. Other stories tell of fishermen trying to trick the selkie into a life on land, where she will spend her days in human form as the mother of the fisherman's children. Otters are also common, but their secretive lifestyle makes them a most elusive animal. They feed mainly on fish, and the size of their territories can range from two to twenty kilometres. The territories of coastal otters are usually smaller due to a better and more varied food supply, while river otters often occupy very long stretches of their chosen waterway. Spread throughout their territory, otters have a number of holts, or underground dens, that they use to rest and rear their young. Despite being seen very rarely, the Irish otter population is one of the most stable in Europe, and the Shannon Estuary is one of their strongholds.

* * *

Human life and wildlife have so far been coexisting in a delicate balance at the Shannon Callows and the Shannon Estuary. While the estuary itself is a busy shipping lane and the shoreline has been fortified and built up for industrial use in many places, there is sufficient room left for fauna and flora to thrive. This, however, doesn't mean that all is perfect. The constant and ever-increasing shipping traffic causes significant noise pollution, which can especially affect the bottlenose dolphins. Industrial waste, untreated sewage and run-off from pastures that can contain chemical elements from pesticides and fertilisers all find their way into the Shannon Estuary, either directly or via its tributaries. Another growing concern is the sheer amount of plastic that is being brought into the estuary through the tidal currents. The upper reach of the tide that was once marked by colourful seaweed accumulations is today a mix of seaweed, plastic bottles and other containers, fishing nets, cans and other flotsam and jetsam of human origin. One of the biggest environmental threats in the area, and a long-running and controversial topic of discussion, is the Aughinish Alumina refinery. Red mud, also known as red sludge or bauxite tailings, is a waste product of the refining process that turns bauxite ore into aluminium. This red mud is highly toxic because of its alkalinity, and at Aughinish it is being stored in large reservoirs close to the estuary. Any leakage could have a devastating effect on all wildlife in the Shannon Estuary. The other two major industrial buildings, the power stations at Killimer and Tarbert, are now in the process of being shut down as part of the climate action plan.

The Shannon Callows are in a similar situation. There have always been calls for flood management, but with the increasing regularity of extreme flood events, these calls grow louder. The ever-growing pressure on farmers to maximise their output also puts the Callows under threat to become just another industrialised pasture, which would mean the end to the rich hay meadows and its birds.

All in all, the future of the Shannon Callows and the Shannon Estuary hangs in an uncertain balance, now more than ever.

Above: Dawn at the Shannon: Fergus Estuary mudflats.

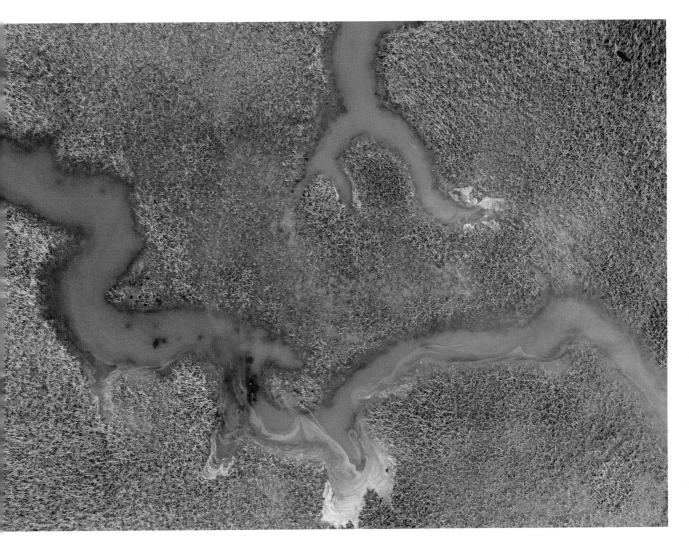

Above: Saltmarsh at Querrin, Shannon Estuary.

Chapter 4

Killarney and the Woodland Heritage

I am lying on the ground, my eyes closed. The dampness of the soft moss blanket I am resting on is slowly creeping through my clothes. I smell wet earth and fresh foliage. A gentle breeze creates rustling sounds in the canopy above. At this early hour, the sun is still hidden behind the peaks of Mangerton and Torc mountains, and the forests of the Killarney National Park lie in deep shadow. The birds, however, are early risers, and the dawn chorus — the symphony that is performed every morning during spring — is about to begin. The voices and the songs of blackbird, robin, wren and others are slowly building. Single performances turn into duets, more singers join in, and soon there is an ensemble of

Opposite: Eagle's Nest, Killarney National Park, County Kerry.
Below: Semi-natural woodland at Dromore Nature Reserve, County Clare.

Above: Owengarriff River, Killarney National Park, County Kerry.

harmonies echoing through the forest. This is the sound of spring: beautiful and uplifting but fleeting.

Over the next few weeks, the birds will establish their territory and find a mate. After they have found their companions, the forest will become a quieter place. The trees will fully unfold their new leaves and the wildflowers on the forest floor will, like the birdsong, fade away. For the moment, however, lesser celandine, primrose, wood anemone, wild garlic and bluebell still add sprinkles of colour to the Killarney woodlands.

Spring will turn into summer and summer into autumn. The fresh and vibrant green that clothed the trees early in the year will turn a darker and duller tone before the forest starts getting ready for winter. In autumn, broadleaf trees stop producing chlorophyll – the pigment responsible for photosynthesis – because over the short days of winter it is more economical for them to rest. The chlorophyll breaks down, and the other pigments present in the leaves – the ones we perceive as autumn colours – are revealed.

In autumn, a different kind of concert is performed in the Killarney woodlands. One October morning, I found myself yet again sitting in the forest in semi-darkness. Suddenly the silence was disrupted by a deep bellowing; it seemed to come right out of the bones of the earth, and in the calm before dawn, it echoed eerily from the mountainsides. The call was answered closer to me by an ear-shattering roar, and I saw a dark shadow moving through the trees ahead. The red-deer rut has a primeval feel; it is an event that has been performed in the forests and around the lakes of Killarney every autumn for thousands of years. Every year, the deer come down from the mountains into the lowlands. Here, the stags round up the does and fight for the right to mate. The bellowing and roaring are followed by violent head-tossing and the crashing of antlers. Often these fights last only a few minutes, and the weaker animal retreats quickly. When two equal opponents meet, however, the conflict becomes violent, and the antlers are turned into deadly weapons. Once, I came across a stag lying in the shelter of a tree, breathing shallowly and bleeding from a wound in its flank. I had witnessed a lengthy fight nearby the day before, and I wondered if this was one of the two opponents. Later, I was told that the injured stag had died from a punctured lung.

In early November, the deer retreat from the valleys, and silence returns to the woodlands once again. It's the calm of the approaching winter. The forest is now fully clad in the warm

Below left: Oak forest in Killarney National Park.
Below right: Reenadinna Wood, Killarney National Park.

colours of autumn, but the spectacle of yellow, orange, brown and red is short lived in this part of the world. The first winter storms can come early and strip the trees bare before they have a chance to change into their full autumn gown. Robbed of their foliage, the masters of the forest appear stark, almost menacing. The bare branches of an old oak tree reaching for the grey winter sky is, for me at least, one of the most awe-inspiring sights in nature and a symbol of utter resilience.

* * *

For the longest time, humanity has admired and even worshipped trees. Their sheer size and longevity make us look at them in wonderment. Trees provide us with food, building materials and fuel. Most importantly, however, today more than ever, they take in carbon dioxide and produce oxygen through photosynthesis. Without trees, life as we know it wouldn't exist.

We treat trees as a commodity, as inanimate objects without consciousness and intelligence. In recent years, however, scientific evidence has confirmed what a small number of naturalists and scientists suspected all along: that trees as well as other members of the plant kingdom are much more than they appear to be. Trees not only actively react to changes and threats in their environment – for example, by releasing foul-tasting chemicals known as tannins into their leaves when someone starts chewing on them – but also warn their nearby companions of the danger. Trees don't necessarily fight for resources but rather share what is available, particularly with their offspring if it is growing nearby. They form communities, not only with other trees but with other plants as well. And they communicate using chemical messaging, both above ground through airborne molecules and underground through their root system, which is connected to a mycelium network that can stretch over vast distances and carry chemical as well as electric impulses. Forest ecologist Suzanne Simard was the first to describe this network as a 'wood wide web', an advanced community of plants, fungi and animals that shares information and food and is not that different from our own towns and cities.

Is this tree behaviour just a genetically programmed reaction to stimuli, or do plants in general (and trees in particular) possess a form of consciousness and the ability to learn? This is an ongoing debate, but it is safe to say that it's time for us to look at trees with a more open mind ... just like our forbearers once did.

In the early 19th Century, forests had almost completely disappeared from Ireland. Back then, only one per cent of the country was covered in trees, with very little old-growth

Opposite: Red deer stag (top) and doe (bottom), Killarney National Park, County Kerry.

Opposite top left: Wood sorrel, Killarney National Park, County Kerry.
Opposite top right: Bluebells, Dromore Nature Reserve, County Clare.
Opposite bottom left: Irish spurge, Glengarriff Nature Reserve, County Cork.
Opposite bottom right: Opposite-leafed golden saxifrage, County Cork.

woodland; the ancient forests that had once covered up to eighty per cent of Ireland's landmass were all but gone.

How it had come to this is hard to understand. In early Irish history, trees were often the spiritual centre for communities, the place to hold rituals, to pass judgement or to crown a new chieftain or king. From the earliest times, trees were treated as sacred, and they were believed to have a deep connection to the Otherworld, the realm of spirits, fairies and other magical beings. Certain trees and shrubs – hawthorn and blackthorn in particular – were seen as a portal to this Otherworld and deemed untouchable, which is reflected in numerous legends and folk tales. These beliefs live on into the present day, and fairy trees and fairy bushes are still very much respected. Not too long ago a section of a planned motorway in County Clare had to be redesigned to circumnavigate a fairy bush, an old hawthorn, that has a firm place in local folklore.

The tradition of the rag tree is related to the fairy tree and fairy bush. Here a piece of cloth taken from the clothing of a sick person is tied to the tree, usually a hawthorn, which often stands close to a holy well, and it is believed that while the piece of cloth rots away so does the ailment of the patient. Today this tradition has somewhat escalated and often photographs, drawings, shoes, toys and other items can be found hanging from trees, and wishes are no longer restricted to the healing of an illness.

Ireland's first written language also has a strong connection with trees. The Ogham alphabet, which was in use from around the 4th to the 8th century, is often referred to as the tree alphabet. It consisted of twenty letters, each made of a combination of short vertical, horizontal and tilted lines. These letters were known as *feda* ('trees') or *nin* ('branches' or 'forks'). It is widely accepted that at least eight of the Ogham letters represented a tree species – the B (*breith*), for example, stood for the birch, the D (*dair*) for

From the earliest times, trees were believed to have a deep connection to the Otherworld

the oak – and that the remaining letters were named with poetic alternatives. The L (*luis*) might have stood for 'flame' to represent fiery red rowan berries, and the T (*tinne*, meaning 'bar of metal or iron') might have stood for the dense, hard wood of the holly.

Dating from the same time period as the Ogham alphabet are the Brehon Laws. This legal

Opposite top left: Puffball, Killarney National Park, County Kerry.
Opposite top right: Amethyst deceiver, Dromore Nature Reserve, County Clare.
Opposite bottom left: Broad-gilled agaric, Burren National Park, County Clare.
Opposite bottom right: Fly agaric, Slieve Bloom Mountains, County Laois.

system, named after wandering lawyers called the brehons, is recognised as the oldest in Europe and remained in place almost unchanged for centuries. Under Brehon law, trees and shrubs were granted special protection according to their importance to the community, and the penalties for breaking these laws were severe. The trees were divided into four classes, roughly mirroring classes in early Irish society: the *airig fedo*, or nobles of the wood (oak, hazel, holly, yew, ash, scots pine and crab apple); the *aithig fedo*, commoners of the wood (blackthorn, elder, spindle, whitebeam, arbutus, aspen, juniper); the *fodla fedo*, lower divisions of the wood (alder, willow, hawthorn, rowan, birch, elm, cherry); and the *losa fedo*, bushes of the wood (bracken, myrtle, furze, bramble, heather, broom, wild rose).

* * *

Remnants of Ireland's ancient forests have only survived in a few small pockets, mainly in the west of Ireland, for example Glengarriff and Uragh in Cork, Derryclare in Galway, Old Head in Mayo and St. John's Wood in Roscommon. The only extensive old-growth woodland is the one around the Killarney lakes, which today forms the heart of Ireland's oldest national park. It was established in 1932 when Arthur Vincent donated the Muckross Estate to the Irish state as a memorial to his late wife Maud and named it the Bourn Vincent Memorial Park. Since then, it has grown considerably, and what is now known as Killarney National Park encompasses an area of over one hundred square kilometres, protecting not only native forests but also lakes, rivers, waterfalls, mountains and areas of blanket bog.

Parts of the Killarney woodlands have been around for some nine thousand years. After the ice sheets of the last glaciation disappeared, the tundra-like landscape was soon colonised by its first trees: birch, willow and hazel. The birch, especially, laid the groundwork for the future forests; its golden leaves, scattered on the ground every autumn, were vital to build up soil. In addition, the birch and some of its relatives like the common alder, also native to Ireland, formed symbiotic relationships with nitrogen-fixing bacteria. These bacteria formed nodules in the trees' root systems, where they produced nitrogen, a vital nutrient. This was made available to the trees and also leaked into the surrounding soil, making the ground more attractive to other plants.

As the soil improved, these pioneer trees — which grew much faster than the big forest trees that were to follow — soon established the first woodlands. A glimpse of what these early forests must have looked like can today be seen in the Burren, a limestone karst area that

Top: Pine marten, Slieve Bloom Mountains, County Laois.
Above left: Red squirrel, Slieve Bloom Mountains, County Laois.
Above right: Jay, Slieve Bloom Mountains, County Laois.

covers parts of counties Clare and Galway. The Burren woodlands consist mainly of hazel with the occasional hawthorn, holly, birch, ash and willow.

After the ground was prepared, the giants of the forest appeared on the scene. Scots pine, elm and oak soon became the dominant species, and the pioneer trees were downgraded to the shrub layer in the new forest hierarchy — or disappeared completely. Only a few birch forests managed to survive in damp corners, on the poor soils of the west coast or at the edge of peatlands. The new kings of the forest, the oak and the elm, dominated the valleys, while Scots pine took over the mountain slopes. Other species like rowan, whitebeam, holly, ivy and honeysuckle thrived in clearings and along lakes and rivers. These primeval forests covered vast areas and were home to a variety of animals. The wolf, brown bear and boar are now extinct, while others like the fox, pine marten and numerous birds are still with us.

The oak woodlands of Killarney, today often referred to as Atlantic Rainforest, are the descendants of these primeval jungles. Oak forests form a thick canopy, and while the trees are in leaf, the forest floor is cast in a permanent twilight. Nevertheless, oak forests support a rich biodiversity. The individual tree is a small ecosystem in itself; several hundred species of plants and animals can live around, in and from an oak. Over two hundred different kinds of insects are known to dwell on a single tree. Bats often use old oaks to roost, and birds like the robin and blue jay like to nest high up in the canopy. In autumn, acorns are an important food source for red squirrels, who pick them directly from the tree, and badgers, who forage the fallen fruit on the ground. Soil can form in nooks and crevices on the oak tree and host wildflowers, while epiphytes like the polypody fern grow directly from the bark, and various mosses, liverworts and lichen are also common hitchhikers. A survey conducted in County Mayo recorded over fifty species of lichen on a single oak tree.

The mild and humid climate of Ireland's west coast is a major contributing factor in making this rich biodiversity possible. Bryophytes — that is, mosses, liverworts and hornworts — particularly cherish the soft weather. A total of 129 of the 162 species known in Ireland thrive in the Killarney woodlands. These forests also have one of the richest collections of ferns in Europe. A total of fifty-two different species grow here, including the royal fern, which can reach heights of up to two metres, and rare species like the Killarney fern, Kerry mousetail fern and Tunbridge filmy fern.

In many places, oaks are joined by plants like the woodrush and bilberry as well as smaller trees like the rowan, holly, hazel, and a special member of the Killarney flora, the strawberry tree. This evergreen species, like other members of the Lusitanian flora, has

Several hundred species of plants and animals can live around, in and from an oak tree

a restricted and rather unexpected range and can only be found on the Iberian Peninsula and the south-west and west of Ireland. Killarney National Park holds the largest strawberry tree community, but the nearby woods of Glengarriff and Derrynane and the forests around Lough Gill in Sligo are also home to this unusual and rare tree. The strawberry tree also has an uncommon fruiting cycle. In September and October, when most other plants carry only fruit, the strawberry tree produces clusters of small, wind-pollinated white flowers. Once pollinated, it takes a whole year for the flowers to turn into the characteristic red fruits that gave the tree its name, and as a result the strawberry tree carries both flowers and fruit at the same time.

The sessile oak found in Killarney is one of the two oak species native to Ireland and overall the more common one. Sessile oak is rather adaptable, and because of this can thrive in fertile valleys as well as on more exposed hillsides with poorer soils. The pedunculate oak, on the other hand, is a bit more demanding and chooses only sheltered sites with nutrient-rich, heavy soils, which can be mainly found in the Irish midlands. The only significant pedunculate oak woodlands are located right in the heart of Ireland, at the Abbey Leix Estate in County Laois and Charleville in County Offaly. The two oak species look very similar and are known to hybridise. The safest way to tell them apart is the acorn: the pedunculate oak grows its acorn on a stalk, while the acorns of the sessile oak are missing this feature. Both species can become true giants — heights of up to forty metres are not unheard of — and reach an age of five hundred years or more. The oldest oak in Ireland, or at least the most likely candidate, is found in the aforementioned Abbey Leix Estate: a pedunculate oak with an estimated age of seven hundred years.

Both species of oak can become true giants and reach an age of five hundred years or more

The majority of Killarney's oaks are younger than two hundred years, and only a few pockets of truly ancient woodland remain. These parts of the forest, like Derrycunnihy Wood, carry a special atmosphere. Gnarled trunks and twisting branches reflect the ballast of the years these beings carry. There is a presence in the air, an authority with the wisdom of centuries that keeps a watchful eye on the place. When walking among these trees, it is easy to believe that the forest itself is very much a conscious entity.

Reenadinna Wood, meanwhile, is very different but demands an equal respect. It stands on an outcrop of limestone that separates Lough Leane from Muckross Lake and is the only surviving yew woodland in Ireland, and one of only a few in Europe. The yew, which grows preferably on limestone and chalk, was among the first trees to arrive in Ireland after the last

Top left: Beech forest, Killarney National Park, County Kerry.

Top right: Scots pines at Lough Leane, Killarney National Park, County Kerry.

Above left: Oaks at Noire Valley, County Kilkenny.

Above right: Portumna Forest Park, County Galway.

Opposite top left: Birch bark, Killarney National Park, County Kerry.
Opposite top right: Ivy, Killarney National Park, County Kerry.
Opposite bottom left: Polypody on oak, Old Head, County Mayo.
Opposite bottom right: Strawberry tree flowers, Killarney National Park, County Kerry.

glaciation but only became widespread around five hundred years ago. Many place names like Mayo (Mhaigh-Eo), the plain of the yews, Drimmo (Droim Eo), the ridge of the yews, or Youghal (Eochaill), yew wood, suggest that it was once far more common than it is today. In Reenadinna, the wild yews clench their roots into the fissured limestone, and the long, low hanging branches create a cave-like space. It feels eerie and comforting in equal measures. This woodland has occupied the stretch of land between the lakes for some four thousand years, and many of the yew trees we see today have been standing for over two hundred. The oldest yews in Ireland are around eight hundred years old and stand at the Crom Estate, County Fermanagh. The Fortingall yew in Scotland, believed to be the oldest in existence, has an estimated age of three thousand years.

* * *

The eventual decline of the old Irish woodlands started with the arrival of the first humans around nine thousand years ago. These early arrivals were Mesolithic hunter-gatherers, and their need for firewood and building material for simple huts would have had only a small impact on the vast forests. The first noteworthy forest clearances happened in the late Mesolithic and early Neolithic periods. At that time, the first farmers had started to create pastures for their animals — cattle, sheep and goats — and ploughed their fields for early variations of wheat and barley. The first evidence of cereal pollen was found some six thousand years ago. At the same time, the amount of tree pollen fell substantially, while the pollen of grassland flora showed a sharp increase.

Next came a change in climate. A relatively dry continental pattern was replaced with the current mild and humid Atlantic regime. This triggered the growth of peatlands, which engulfed vast areas of woodland, especially in the west of Ireland. The Scots pine was one of the first victims of this event, and despite some temporary recolonisation of the young peatlands during drier intervals, the Scots pine had disappeared from the country by the early Bronze Age. Petrified tree trunks embedded in the layers of peat are a striking reminder of this period. The elm, which, together with the oak, had been one of the main species of Ireland's early forests, also disappeared completely. It is thought that the combination of farming, where elm foliage was used as cattle fodder, and a disease similar to today's Dutch elm disease led to the disappearance of this mighty forest tree from Ireland by the 7th Century.

A similar malady is currently threatening the ash, a tree that is quintessentially Irish in many ways. The ash is not only a native and an integral part of woodlands and hedgerows, it also traditionally provided the wood for the hurley, the stick used in hurling and camogie. Ash dieback disease is caused by *Hymenoscyphus fraxineus*, a fungus that spreads through the air and eventually kills most of the trees it infects. Ash dieback was first reported in Poland in 1992, and after making its way through Europe, the fungus arrived in Ireland in 2012. *Hymenoscyphus fraxineus* is not native to Europe and was brought to Poland from warmer climates, probably on the shoes of a traveller or stuck to a shipping container. The milder temperatures caused by climate change allowed the spores to survive and seek out their first European victims. Because the European ash had never experienced *Hymenoscyphus fraxineus* before, the trees had no defence mechanism and became easy prey for the fungus. Whether or not the ash as a species will survive is unknown. Some trees reportedly withstand the infection for longer than others, but a complete immunity or recovery hasn't been seen yet. In the best-case scenario, the ash will develop defences against the contagion and make a comeback; in the worst case, the tree will disappear from the Irish landscape.

* * *

In a 9th Century text, only a few forested wilderness areas are mentioned: Fid Mar hi Cuailngi (the great wood in Cooley, County Louth), Fid Deicsen hi Tuirtre (the wood of Deicsiu in Tuirtre, around Slieve Gallion, County Tyrone) and Fid Moithre hi Connachtaib (the wood of Moithre in Connacht). Ireland in general is described as a landscape of fields interspersed with single trees, rows of trees and small woodlands. Pollen records support these descriptions, and it seems that Ireland had already lost most of its forests before the Norman invasion in the 12th Century. This, however, doesn't mean that the new ruling class took particular care of Ireland's remaining woodlands. Forests were treated as a cheap resource, there for the taking, and a substantial amount of Irish timber made its way to England, where it was used in all kinds of building projects, from houses to ships. It is estimated that by 1600 only twelve per cent of Ireland was covered in forests, and continued exploitation brought this down to around two per cent by 1800.

During the 16th and 17th centuries, new species were introduced into the country estates and formal gardens that had become a regular feature of the Irish countryside. Among them were beech, horse chestnut and sycamore, which have all become a common sight in Ireland. The beech has established itself as a forest tree. In Killarney, it thrives in particular on the Ross Peninsula, where it displays a typical vibrance with its wonderful combination of light-green foliage on the trees and a cover of last year's golden leaves on the ground.

Another introduced species is the notorious rhododendron. In contrast to its spectacular flower display, this evergreen shrub is a constant threat to the native flora. The rhododendron responds extremely well to the mild and damp climate and can take over vast areas very quickly. Because of its dense growth, it kills off any competing plants quickly and effectively. The National Parks & Wildlife Service, with the help of volunteers, is fighting a never-ending battle against the rhododendron, not only in Killarney but all over Ireland.

The difference between native and introduced, good and bad, is not always clear. The Norway spruce, for example, didn't make it into Ireland after the last glaciation and was recently introduced as a plantation crop, mainly for Christmas trees. Cones found in Antrim and Mayo, however, show that the tree was growing in Ireland during an earlier interglacial period some ninety thousand years ago. This raises the question of what 'native' really means. For how long does a species have to be gone from Ireland to become 'non-native'? And how long does it have to be growing or living in Ireland to become 'native'?

✳ ✳ ✳

Undoubtedly native is the pine marten, Ireland's much-loved forest predator. The slender, elegant mammal is not a pure carnivore but also enjoys nuts and berries – as well as the occasional chocolate bar. The latter I found out during my first encounter with this elusive animal. I had left a half-eaten Snickers bar in my open backpack, which was sitting a good few metres away from where I had set up my camera. I was immersed in my work when a grunting and rustling caught my attention. I turned around, only to see the backside of a pine marten sticking out of my bag. I guess I must have gasped or made some other sound of surprise, because the marten turned around, the rest of my Snickers in its mouth, stared at me for a moment, and then took off.

By the early 20th Century, the pine marten was almost extinct. A combination of being hunted for its fur and the almost complete disappearance of its habitat reduced the population to only a few animals in a few secluded places along Ireland's west coast. The introduction of commercial forest plantations some decades later came just in time. While these monocultures rightly have a bad reputation, they very likely saved the pine marten from extinction. The plantations provided shelter and created stepping stones in the landscape, and slowly but surely the number of pine martens started to climb and spread around the country.

Around the same time the grey squirrel, a North American native, made its way into Ireland. Because of its bigger size, faster breeding cycle and more aggressive attitude, it started to replace the native red squirrel, especially along the east coast and in the midlands. In the early years of the 21st Century, conservationists feared the worst for the future of the cute, rust-coloured mammal. A survey conducted in 2007, however, showed a surprising development: grey squirrel numbers were in decline across the country, and reds were making a comeback. The most likely explanation for this unexpected situation is the growing pine marten population. Red squirrels are the main prey for the marten, but over millennia they have developed strategies to avoid or escape the hunter. To the grey squirrel, on the other hand, pine martens are a new and unfamiliar threat, which makes them an easy meal. It seems the Irish pine marten population has taken advantage of that.

In the early 20th Century, a slow process of forest recovery began. Since then, afforestation, forest management and conservation measures have increased Ireland's forest cover to around eleven per cent, which is still the lowest in Europe, where the average lies around thirty per cent. Most of these new forests are commercial conifer plantations made up mainly of fast-growing and non-native species like the notorious Sitka spruce. Only around twenty-five per cent of Ireland's current forests consist of broadleaves, and native trees like oak, ash, birch and hazel make up only half of that percentage.

As we are travelling deeper into the 21st Century and the impact of climate change and biodiversity loss is starting to show, the attitude towards forests in Ireland is changing, and huge efforts are underway to increase the country's forest cover quickly and permanently. Numerous private initiatives concentrate on establishing new and enlarging existing woodlands by planting native broadleaf trees like oak, ash and birch.

The previous afforestation attempts don't have a long-lasting positive impact on the natural world because, like any crop-plant, trees are being harvested once they have reached maturity, and what is left behind is nothing but a wasteland. To make a real difference, a long-term approach is needed. This means planning and managing woodlands in a way that allows continuous regeneration, while at the same time carefully harvesting trees for commercial use. The solution is mixed woodlands with a combination of fast- and slow-growing species, the reintroduction of traditional forestry methods like coppicing, and ongoing replanting. This will create forests that will exist for centuries, produce a regular crop and at the same time host a rich plant and animal community.

There is ongoing debate about how to create these forests — whether to methodically plant them or let them develop naturally. Some environmental organisations suggest that we should allow at least some woodlands to grow naturally, without human interference. These forests wouldn't be used commercially but would be dedicated natural havens where wildlife could thrive without any disturbance from outside. The Burren National Park in County Clare is one of the places where this natural development can be witnessed in real time. To conserve the unique wildflowers of this karst landscape, however, the Burren is currently being actively managed. Traditional farming techniques keep developing woodland in check, and encroaching scrub is cut back on a regular basis. If we allowed these shrubs to prosper, the series of events that took place after the last glaciation would be repeated: hazel, blackthorn, hawthorn and other shrubs would be followed by bigger trees, and over time the Burren would revert to a mixed forest. In some spots of the Burren National Park, this process is being allowed to happen and ash and birch are already growing among the hazel.

Whether or not this approach would really work is questioned by a long-term experiment that was started at Lady Park Wood, a forested area at the border between England and Wales, in 1944. While some parts of the forest have developed an extremely rich and healthy biodiversity, other parts have gone the other way. Because Lady Park Wood didn't exist in a bubble, separated from the real world, invasive species still found their way in, and unchecked grazers, because of the lack of natural predators, also did considerable damage. In order to create a truly natural woodland, we would first have to eradicate all invasive species like rhododendron and the sika deer and reintroduce at least some extinct native species, especially predators like the wolf.

I have been living in a remote, treeless corner of the west coast of Ireland for over twenty years now. As much as I love the ocean and the wide-open skies, for someone who grew up in Germany, not too far away from the Bavarian Forest, and lived a decade near the Black Forest, the lack of trees is hard to bear, and at times I long for the embrace of the forest and the smell of damp leaves and moss. Studies have shown that spending time in the forest has not only a positive effect on our mental health but also results in physiological improvements like reduced blood pressure and a boost of autonomic and immune functions. Our species has been living in and from the forest for millennia, and only over the past centuries have we turned away from our ancestral green home. The heritage, however, is inside us. Somewhere deep down, we are all still forest people.

To create a truly natural woodland, we would have to reintroduce at least some extinct native species, especially predators like the wolf

Above: Dawn at Upper Lake, Killarney National Park, County Kerry.

Opposite: Marsh marigold in wet woodland, County Laois.

Above: Beach and oak forest at Old Head, County Mayo.

Above: Upper Lake, Glendalough, Wicklow Mountains National Park, County Wicklow.

Chapter 5

Fields and Hedgerows

The weather forecast promised a sunny spring morning, so I set my alarm for 6am to take a walk at first light along 'my' local boreen. This boreen is not, as the name implies, a small country road but rather a narrow and rough farm track that gives access to a number of fields. Its southern side is marked by an old dry-stone wall, which is periodically interrupted by metal gates. In parts, the stone walls have completely disappeared under an array of grasses and wildflowers that provide ever-changing colours throughout spring and summer. Where the wall still shows through, it displays small stones – pebbles almost – dug out of the soil and then painstakingly stacked up to a height of about a metre. Most of the stones are covered in lichen, which forms intricate patterns of brown, grey and white, and in the narrow crevices between the rocks, ferns and mosses have found the sheltered dampness they like.

It is impossible to determine what the northern boundary of the boreen was originally made of. Whether it was another stone wall or an earthen bank, it has completely disappeared under a concoction of grasses, bramble, gorse, bracken, heather and other wildflowers. In parts, this green wall reaches up to two metres in height, and here live the

Opposite: Farming landscape, County Louth.
Below left: Female stonechat, County Clare.
Below right: Wren, County Clare.

Opposite: Blue tit, County Clare.

feathered individuals I was hoping to see this morning. The weather had turned out as promised, and I enter my boreen just as the sun sends its first rays over the horizon.

The wren is the first of the resident birds to appear. Sitting on an elevated bramble shoot, it chants its characteristic — and for a bird this small, very loud — song. This is, however, not the bird I'm looking for. I walk deeper into the world of the boreen. With every step, I get further away from the domesticated farming landscape of fields and pastures and into a place that feels pristine and wild. Only the mooing from the other side of the hedge and the sound of a distant tractor retain a touch of reality. Then I hear the song I am waiting for, the sound of pebbles being banged together, the voice of the stonechat. At least two couples of this common countryside bird are regularly breeding in the thickets of the boreen, and spring is the best time to photograph them. At this time of the year, they are extremely protective of their territory, and any intruder will be approached and called out. The female stonechat is sitting down on a newly emerged bracken frond, eyeing me suspiciously, while the colourful male is perched a bit further away, continuing his warning song. A few metres away, a blue tit appears on top of a gorse bush, apparently checking what the ruckus is all about, and a blackbird rises out of the brambles, shouting its own alarm call while flying away.

I spend the better part of the morning in my boreen bubble, watching the birds that depend on this limited wilderness. I also discover the unrolling fronds of the heart's tongue fern, the blue flowers of the common dog violet, the yellow primroses, and the crimson carpets of English stonecrop that later in the year will produce precious white flowers. This boreen is only a few hundred metres long and less than five metres wide, but for its inhabitants, it is the whole world.

* * *

An aerial view of Ireland reveals the characteristic checkerboard landscape this island is so famous for. Small fields in various shades of green are separated by rows of shrubs and trees or overgrown embankments and stone walls. Over the centuries, and especially over the past few decades, these dividers of the agricultural landscape have become one of the most important habitats for wildlife in Ireland. While fields and pastures are no longer able to support a wide variety of species due to the ever-intensifying monocultures that come with modern farming, hedgerows are now islands of biodiversity in a sea of bleak pastures and fields.

The origins of Ireland's checkerboard landscape date back to Neolithic times. The oldest known cultural land in Ireland sits on a gentle slope overlooking the Atlantic Ocean on the

Opposite top left: English stonecrop, County Clare.
Opposite top right: Purple loosestrife, County Clare.
Opposite bottom left: Lichen on stone wall, County Galway.
Opposite bottom right: Buttercups and herb Robert in the herbaceous strip, County Galway.

north Mayo coast, just west of Ballycastle. The Céide Fields were discovered in the early 20th Century by a local schoolteacher while he was cutting his turf for the winter, and subsequent excavations have revealed a vast network of fields separated by stone walls, several dwelling houses and megalithic tombs. It was the remains of a Neolithic farming community, preserved in the peat.

The tradition of using stone walls as a border between fields has lived on into the present day. The west of Ireland, in particular, is hard to imagine without its endless lines of dry-stone walls. The use of these structures as a field divider was very likely born out of necessity. Soils along the west coast are generally poor, and the first action that would have been taken to create a usable pasture was removing the stones lodged in the dirt. These stones were then used to build walls around the small fields to keep livestock in, and to keep the ever-present wind from removing what little arable soil there was.

Building these walls has become a craft, even an art form, in certain areas. In the limestone landscape of the Burren, on the Aran Islands and in parts of Connemara, this craft has reached an unmatched sophistication, and walls are kept in pristine condition. In parts of the country where stone walls have fallen into disuse, meanwhile, they have been fully or in parts claimed back by nature, becoming a habitat in their own right. Soil that accumulated in the crevices between the stones is hosting ferns and wildflowers like polypody, navelwort, English stonecrop and ivy-leaved toadflax. Near the coast, thrift, sheep's-bit and lesser hawkbit have colonised and in places completely covered the walls. The stones themselves are almost always covered with lichen and mosses, and the nooks and crevices give shelter to the pygmy shrew, the field mouse, the bank vole, the viviparous lizard and other small animals. Insects, in particular, depend on this habitat not only for shelter but also for food. The rare great yellow bumblebee and shrill carder bee have often been recorded near old stone walls, and in 2021 the thrift clearwing moth — thought to be extinct — was rediscovered in overgrown stone walls on the Loop Head Peninsula in County Clare.

Away from the exposed west coast and predominantly in the midlands and the east, hedgerows replace the stone walls. When this well-known feature of the Irish landscape became first established is not entirely clear, but it is likely that hedgerows also date back to the early days of

Building dry-stone walls has become a craft, even an art form, in certain areas

farming. Ireland's present hedgerows were planted over the past few centuries, so we can only guess how the very first came into being and what they looked like. Even today there is considerable similarity between the species that make up the hedgerows and those of the nearby woodlands, which leads us to ask if the first hedgerows were left standing deliberately or by chance when early farmers cleared the woodland, or if they were planted later using locally available species.

What is known for certain is that building field enclosures was common practice by the early years of the first millennium. Texts from the 7th and 8th centuries describe temporary

Below left: Hedgerows, County Meath.
Below right: Hedgerows at sunrise, County Down.
Bottom left: Boreen, County Clare.
Bottom right: Hedgerows and pastures, County Cavan.

solutions like movable panels of wattle – a lattice made from strips of wood – but also earth walls topped with branches of blackthorn. It is likely that some of these blackthorn cuttings would have taken root; other flora would have invaded subsequently, and a hedgerow was established.

By medieval times, the hedgerow had become much more than just a fence to mark field boundaries and separate pastureland from tillage and livestock from crops – deliberate planning and laying of hedgerows was very much part of the farming regime. The hedgerow itself had become a component of the farm yield, a source of food as well as timber, which was used for fuel and as a building material. In addition to the primary hedging plants, mostly blackthorn and hawthorn, other plants like crab apple, wild plum, wild cherry and hazel had been introduced. Hazel was particularly popular for its versatility. Not only did it provide high-energy food in the form of nuts, its stems were used to make wattle and baskets. They were also favoured by water diviners for their allegedly magical properties, and by pilgrims as a walking stick.

Today some 830,000 kilometres of hedgerow criss-cross the Irish landscape, and these checkerboard lines have become a refuge for flora and fauna. At present, they are more important for biodiversity than any other terrestrial habitat. On average some one hundred different plants make up a hedgerow, but the composition of species varies considerably and is dependent on location, soil type, and how, if at all, the hedge is being managed.

* * *

The blackthorn – despite its mostly sinister reputation in Irish folklore, where it appears as an ill omen and in connection with warfare and death – is still one of the most common and characteristic plants in the Irish landscape and one of the two main building blocks of a hedgerow. It is easy to distinguish from other shrubs as it is the first to come into bloom, sometimes as early as March, and the only shrub that carries flowers before it produces leaves. These delicate white flowers, which stand in stark contrast to the almost black branches, turn into perfectly round, succulent berries known as sloes that shimmer in a dark-blue colour in the autumn sunlight and are traditionally used to make gin.

The second main building block of a hedge is the hawthorn. Like the blackthorn, this shrub grows quickly and can tolerate a variety of soil types, making it the perfect candidate to establish a new hedge. Hawthorn produces distinctive clusters of flowers that stand out from the dark-green foliage in May and June. These flowers are white, or white with a soft, pinkish hue, and on closer inspection reveal delicate filaments topped by anthers that turn from a bright pinkish red to dark brown over the lifetime of the flower. These flowers, together with the young leaves, were once a popular snack known as 'bread and cheese'.

Opposite top left: Blackthorn, County Galway.
Opposite top right: Hawthorn, County Donegal.
Opposite bottom left: European gorse, County Clare.
Opposite bottom right: Ivy flowers, County Clare.

Over the summer, the flowers transform into small, crimson berries which can be made into a wonderful jam or jelly.

Over time, blackthorn, hawthorn and other deliberately planted hedging plants would be joined by wild species: spindle tree and guelder rose are common in the lime-rich soils of sheltered lowland sites. Both develop clusters of exquisite, small white flowers in early summer, but it is in autumn when these two shrubs are at their most striking. From September, they display glowing tones of pink, red, orange, yellow and brown in both their foliage and fruit.

Gorse is not a typical hedgerow plant, but on the poorer soils of the uplands and exposed coastal sites, it can be seen on its own or as part of a hedge. The European gorse, also known as furze or whin, is without a doubt one of the most striking plants in Ireland. Its leaves have been transformed into spines to reduce water evaporation and as a defence against grazing animals. When not in bloom, this shrub is nothing much to look at and wears a dull green coat. In spring, however, and at times again in autumn, the gorse produces countless yellow flowers. The flowering of the gorse is one of the major events in spring, and seeing hedgerows, mountainsides and coastal slopes transformed and brightened up by a sea of yellow flowers is truly uplifting after a long Irish winter.

Today, gorse is often seen as a nuisance, but only a century ago the shrub was used as protective fencing, for harrowing fields, as a chimney brush and as fuel. Bakers particularly appreciated gorse wood due to its high oil content, which allowed for a long and hot burning flame. This characteristic unfortunately also makes gorse susceptible to wildfires.

A close relative of the European gorse is the Western gorse. This shrub is considerably smaller and appears more delicate than its more common relative. Western gorse also prefers higher altitudes and only very rarely appears in hedgerows. The most obvious difference between the two species, however, is their smell — or lack thereof. The flowers of the Western gorse are completely odourless, while the European gorse exudes an overpowering coconut aroma.

Another well-known, almost iconic, shrub of Ireland's hedges is the fuchsia, with its unique bell-shaped red and purple flowers. The plant is particularly associated with the west coast but can be found all over the country.

The flowering of the gorse is one of the major events of spring and truly uplifting after a long Irish winter

Opposite top left: Guelder rose, County Clare.
Opposite top right: Spindle, County Clare.
Opposite bottom left: Field rose haws, County Kerry.
Opposite bottom right: Dog rose, County Clare.

Unlike the plants described so far, fuchsia, which was named after the German botanist Leonhart Fuchs, isn't a native but was introduced from South America as a garden plant and eventually escaped into the wild.

Once a hedge is established and reaches a proper height, other plants, first and foremost the climbers, arrive. The most common of these is the bramble. This hardy species comes in more than eighty microspecies, which often appear together in the same hedge. The difference between these microspecies is almost impossible to tell for most of the year, but it becomes somewhat apparent in autumn when the blackberries ripen. Botanically speaking, the blackberry is not a berry but an aggregate fruit, consisting of multiple drupelets, each of which contains a seed. Some 'berries' are made up of only a few loosely arranged drupelets, while others consist of many, tightly packed. Some berries are small and hard, others are big and juicy. Some exude a sweet smell, some don't. Some are ready to eat in August, others need until early October before they have turned from green to the characteristic deep, dark blue. All these different fruits represent a different microspecies, but what benefit these variations bring to the individual plant is unknown.

Where brambles thrive, often honeysuckle isn't far away. The flowers of this plant are unlike any other. All parts, from the petals to the style, are slender, graceful and coloured in soft, creamy colours. The British poet Lord Alfred Tennyson was one of many enchanted by this lovely plant and wrote in 1859 of 'how sweetly smells the honeysuckle in the hush'd night'. The scent that so captivated Tennyson is aimed at the insects who are the main pollinators of the honeysuckle: bumblebees during the day and hawkmoths at night.

Ivy and members of the rose family can also find their place in the hedgerow. Ireland is home to nine or so species of wild rose. The problem with pinpointing a species is that all wild roses tend to hybridise, which results in a considerable number of microspecies that even experienced botanists have a problem telling apart. The dog rose, with white and pink flowers, the field rose, with pure white flowers, the sweet briar, with mostly pink flowers, and any hybridisation thereof, are the most common roses of the hedgerow.

Ivy plays a very special and often underestimated part in the yearly cycle of the hedgerow. Unlike other members of the community, ivy is not only an evergreen plant, it also starts to bloom when all the others get ready to retire for winter. Ivy produces its flowers in September and October and is therefore a vital food source for insects stocking up their reserves before hibernation. A few months later, in January and February, the dark-blue, almost black, berries of the ivy help birds to get over the worst part of the winter.

Some hedgerows also feature proper trees like rowan, sycamore, elder, beech or oak in addition to the various shrubs and climbers. These trees were either planted to fortify the hedge or they invaded the hedge naturally, as would happen during the development of a new woodland.

* * *

Hedgerows and stone walls rarely stand on their own. The fields and pastures they encircle have their origin in Neolithic times, when farming started to replace hunting and gathering as a way to make a living. The semi-natural grasslands created by the early farmers were made up of some 250 species of flowering plants. These species formed communities depending on local climate, soil type and drainage condition, and it wouldn't have been unusual to find more than forty species within a few square metres. Today these grasslands are divided into four main categories: wet meadows, dry meadows, roadside grassland and upland meadows. These environments often blend together, sometimes even in the same field, and so can host a multitude of different species. The majority of the angiosperms in a meadow are grasses. Worldwide there are more than 11,500 known species of grasses, including our cultivated food staples of wheat, rye, oat, corn and rice, and agrostologists (those who study grasses) estimate that there could be a total of up to thirteen thousand different species. One of the most common species in Ireland is red fescue, which can tolerate a range of soil conditions. Marsh foxtail is an indicator for wet ground, quaking grass prefers it well drained and dry, sweet vernal grass is typical for acidic soils, and yellow oat grass is an indicator for alkaline soils.

The wind-pollinated grasses grow side by side with a variety of meadow flowers like common knapweed, oxeye daisy, red and white clover, bird's foot trefoil, ragged robin, devil's bit scabious, various orchids, and others. These grasses and herbs thrived not despite but because of the interference of farming. Carefully timed grazing and cutting allowed the herbaceous species to flourish beside the grasses, and the droppings of farm animals provided fertiliser and supported the regeneration of the soil. This way of farming very much reflected the natural way of things that allows a wide variety of plants and animals to thrive.

The best-known of the semi-natural grasslands is the traditional hay meadow, a dry grassland that can be found on calcareous and well-drained soils. Its counterpart, the wet meadow, often lies on the floodplain of a river or beside a lake that experiences fluctuating water levels. The meadows of the Burren National Park are a good example of the former; the Shannon Callows, of the latter. A unique kind of grassland and one of the rarest habitats in Europe, only found in the west of Ireland and parts of western Scotland, is the machair. Machair is the Gaelic word for 'fertile plain', and these low-lying coastal grasslands are always adjacent to a beach or dune system. The ground that forms the machair is highly calcareous and consists mainly of shell fragments that have made their way from the sea onto the beach

Top left: Digger wasp, County Clare.

Top right: Lackey moth caterpillar, County Clare.

Above left: Hover fly, County Clare.

Above right: Yellow dung fly, County Clare.

Opposite: Pyramidal orchid in machair, Connemara, County Galway.

and eventually inland, being broken down into ever-smaller particles along the way. In summer, the machair turns into a colourful carpet of pyramidal orchid, oxeye daisy, bird's-foot trefoil, lady's bedstraw, white clover, harebell and others, and it rivals any hay meadow in appearance.

While these grasslands were never truly natural, they formed a species-rich environment where various plants, insects and birds existed in relative harmony and balance with men. Today, unfortunately, these grasslands have mostly disappeared from Ireland. The old farming practice of haymaking — which isn't, in fact, all that old and only came into being around a thousand years ago — involved one cutting in late summer and tedious stacking and drying of the bounty. This labour-intensive and, in the wet Irish climate, often futile practice has been replaced by silage, the cutting and immediate wrapping of the crop into plastic sheets. The new technique is a much safer and quicker option for the farmer to secure winter fodder for the animals, but the arrival of silage was also the start of a downward spiral for the biodiversity of the species-rich grasslands. A growing national herd demanded more food, and so the old meadows with their mixture of grasses and herbs were replaced with fast-growing rye grasses. At the same time, chemical fertilisation was introduced to allow two or even three cuts over the summer. We are now left with beautifully green but otherwise barren pastures; where once thirty or more species thrived in a field, today we would be lucky to find ten.

The loss of plant diversity is, however, only the tip of the proverbial iceberg. Apart from the wrapping plastic that has turned into a pollution problem in itself and the run-off of fertiliser that causes additional environmental problems, modern farming has also displaced many ground-nesting birds. The nests and their inhabitants that aren't being destroyed by farm machinery are an easy pick for predators. The corncrake is probably the most famous (but by no means the only) victim of modern farming. It has disappeared from most of its traditional breeding grounds and is close to extinction. Another and more far-reaching side effect of modern monocultures and the use of chemicals is the decline in insect populations, and therefore the loss of pollinators. In recent years, not only have our much-loved bees and butterflies dramatically dropped in numbers, the not-so-much-loved but nevertheless important flies and other creepy-crawlies are in sharp decline as well.

The corncrake is probably the most famous (but by no means the only) victim of modern farming

The original meadow flora, however, manages to survive in many places in a narrow strip that runs along the field boundary, which is known as the herbaceous strip. Together with drains and ditches, built to prevent the fields from flooding, the hedgerow, stone wall and the herbaceous strip form one of the most species-rich and diverse habitats in Ireland.

Drains and ditches can often display plant communities similar to those along lakes and rivers, with reeds growing at the edges and water cress, marsh cinquefoil, water mint and other lovers of a wet environment flowering in and around the trench. Only a step away from this miniature wetland, you will find flowers of calcareous grassland, and another step will bring you to plant communities of a woodland edge.

Because hedgerows very often resemble woodland edges, the fauna found in both habitats is similar too. Ireland's smallest mammal, the pygmy shrew, which reaches a length of only five centimetres, the field mouse and the hedgehog are typical hedgerow dwellers. The hedgehog can be found noisily foraging the damp soil under the branches for worms and other delicacies, hence its name. The stoat, fox and badger are also regular inhabitants and use the hedge for both temporary shelter and a home for raising their offspring.

The hedgerow is of particular importance for birds. In many areas, it provides the only suitable place for finches, thrushes, tits, flycatchers and tiny wrens to build nests and rear their chicks. Surveys have shown that around fifty per cent of all recorded countryside birds use hedgerows more or less exclusively as their homes, and that the majority of these birds prefer hedges of a certain size. A height and width of over one metre seems to be the sweet spot between accessibility and shelter from ground and aerial predators like the fox, the stoat, the invasive American mink, the sparrowhawk, kestrel, long-eared owl and barn owl.

The barn owl is a typical hunter of the woodland edge. It needs trees or shrubs on which to perch and seek out the next meal, and the open space of an adjoining field to silently swoop down for the attack. Like most owls, the barn owl is a night and twilight hunter. Despite its big eyes, it doesn't rely on eyesight to locate and target its prey; instead, it hunts by sound. A barn owl's ear openings are enormous in relation to the size of the head. In addition, one ear sits slightly higher than the other, which allows the animal to pinpoint the source of a sound with greater accuracy. The barn owl's secret weapon, however, is the characteristic shape of its face. The flat facial disc and the ring of special feathers around it acts like a satellite dish, capturing the sound of even the slightest movement on the ground beneath the perch. Unfortunately, the barn

The barn owl's secret weapon is the characteristic shape of its face

owl is on the brink of extinction due to habitat loss; numbers are steadily declining, not only in Ireland but across the whole of Europe.

For many animals, the hedgerow is not only a place to dwell, raise their offspring, forage or hunt, it is also a road network. Travelling over open ground runs the risk of being spotted by the wrong pair of eyes, and this applies to all animals – mammals, birds, insects and other invertebrates. The hedgerow provides much-needed protection while getting from one place to another.

In addition to their importance for plants and animals, hedgerows fulfil even more far-reaching services. They control the water saturation of the ground and can avert flooding. Their root systems prevent erosion, and their foliage can cool down air temperatures through evaporation, thereby influencing local climate. All these factors have a positive effect not only on the natural biodiversity but also on farm animals and crops.

Unfortunately, hedgerows and stone walls don't go well with modern farming practices. The trend is for bigger farms and bigger herds, which need bigger fields and bigger pastures. Where partitions are needed, wire fences are easier to install and maintain. The disappearance of these old structures is not only a loss for nature, it is also a loss of heritage and local identity that transforms the countryside into an ever more uniform and sterile place. The diverse farming landscape of old is becoming an empty factory floor. The improper maintenance of hedges, especially the indifferent hedge-cutting performed by large machinery, also causes considerable damage and destroys natural communities.

How Ireland manages its hedgerows, fields and pastures in the future will have a massive impact on the country's biodiversity and the effects of climate change, and, subsequently, on the livelihood of farmers.

Opposite: House sparrow feasting on dandelion, County Clare.

Top: Fox, County Clare.

Above: Male stonechat at dawn, County Clare.

Chapter 6

Wonderful Peatlands

One of my earliest memories of Ireland is standing in the narrow streets of Dingle in the twilight of a cold and damp winter evening, the air heavy with the smoky-sweet smell of turf fires. A biting wind was blowing from the Atlantic, and there even might have been some snowflakes in the air. To this day, the unique aroma of smouldering peat encapsulates everything Ireland means to me. It's the smell of a cosy seat by the fire, the smell of shelter, the smell of home.

Some years after this magic moment in Dingle, I had settled in Ireland and was living in County Clare, in an old cottage in the middle of the bog. My kitchen window looked out on a wide stretch of cutover Atlantic blanket bog and wet pastures, giving me a front-row seat from which to learn the sights and sounds of the peatlands. In spring, when fresh green was sprouting out of the brown peat carpet, the Irish hare would throw all caution to the wind. Groups of these usually elusive animals chased each other across the fields and along the old bog road that gives access to the peat banks. In summer, the song of the skylark rang out over a sea of vibrant green and pure white made of sedges and cotton grasses. Between the songs, a blanket of stillness lay over the landscape. Midges danced silently in the shimmering heat, while a dragonfly hung apparently motionless in mid-air and a butterfly nipped at the flowers of the heather. In late August, the rain returned, replenishing the ground and the dried-out, almost white sphagnum mosses. In autumn, when the bog had stripped off its summer gown and now glowed in warm tones of brown and yellow, the heart-breaking cry of the curlew described the coming dark days of winter, and I watched lapwings and brent geese in the fields while washing the dishes.

Back then, I was only starting to learn the secrets of the bog, the connection between men and peat, and the journey the dark-brown briquettes undergo before ending up in a fireplace. I didn't know about the days, weeks and months that it takes to turn the squelchy peat into turf, the wielding of the *sleán* to cut the sods, and the backbreaking endeavour of turning and footing them. I wasn't aware of the thousands of years of growth and death, the layers of history and the stories that are wrapped up in the peat. Most of all, I hadn't seen yet what industrial turf cutting had done to the peatlands. I hadn't felt the sense of loss when looking over the brown, barren wasteland that had once been the thriving Bog of Allen, a raised bog stretching out over five counties in Ireland's midlands.

Opposite: Cutover Atlantic blanket bog, Connemara, County Galway.

* * *

For the past four thousand years, Ireland has been dominated by peatlands. These delicate habitats stretch from the coastal fringes to the highest mountains and cover the wide expanse of the Irish midlands. It is a landscape of understated yet mesmerising beauty, rich in fascinating plant life and home to some wonderful animals. On a global scale, some five million square kilometres of peatland cover the earth's surface, from the tundra and permafrost regions of northern Europe, Russia and Canada all the way to the southern hemisphere and places like Chile, Indonesia and Africa's Congo Basin.

Peat, the unique brown and squelchy substance that makes up the bogs, consists mainly of dead plants that failed to decompose due to the absence of microorganisms. However, not all peat is the same. The two main variations of peatlands in Ireland are raised bogs and blanket bogs. The latter can be further subdivided into Atlantic (or lowland) blanket bogs and mountain blanket bogs. The lowland blanket bog stretches from sea level to around 150 metres, where it starts to merge with the mountain blanket bog, which reaches its borders at around 300 metres. The main difference between the two types is their vegetation. Lowland blanket bogs are dominated by purple moor grass and black bog rush, while the slightly drier mountain blanket bog is ruled by ling and hare's tail cotton grass.

The peat of raised bogs and blanket bogs differs significantly in both colour and consistency. Peat from raised bogs is mainly made of sphagnum mosses and is rather light in colour, while peat from blanket bogs shows a very dark brown colouring and consists mainly of grasses and sedges.

The formation of raised bogs started at the end of the last glaciation, some ten thousand years ago, when the central plain of Ireland was dotted with countless shallow lakes and ponds left behind by the retreating ice sheets. As temperatures rose, reeds and sedges developed around the edges of those bodies of water, and floating plant communities thrived on the water surface. Because of their glacial origin, these lakes supported only small numbers of microorganisms like bacteria and fungi, which are vital for decomposition. As a result, the dead members of the plant communities didn't decompose completely but started to accumulate at the bottom and around the edges of the lakes. While Ireland experienced a more continental climate, this was a very slow process, but when the climate changed to the much wetter and milder Atlantic regime of today, plant growth increased and decomposition ceased almost completely. The layers of dead plant material, known as fen peat, grew quickly

Opposite top left: Clara Bog, County Offaly.
Opposite top right: Atlantic blanket bog at Ballycroy, Wild Nephin National Park, County Mayo.
Opposite bottom left and right: Raised bog landscape at Clara Bog, County Offaly.

Above left: Summer morning in the bog, County Clare.
Above right: Lough Boora, County Offaly.

from the edges and the bottom of the lake. This allowed the sedges and reeds to expand their reach until eventually the whole lake was colonised. Over time, other plants moved in, and at this stage, a fen was formed.

What differentiates a fen from a bog is a constant water supply other than precipitation. This water supply can be one or more underground springs or a river, all of which provide a range of vital minerals that support a wide variety of plant life. Fens are constantly changing and therefore aren't a uniform landscape. A typical fen is a kaleidoscope of small areas of open water, stretches of marsh, and wet woodland known as fen carr. A fen can support up to two hundred species of plants, from grasses and sedges — which alone are represented by some fifteen different species — to trees like willow, alder and birch. Numerous wildflowers also thrive here: the marsh cinquefoil, devil's bit scabious, meadowsweet, lady's smock, marsh orchids, butterfly orchids and many others. This flora, in turn, supports a variety of wildlife. Insects and other invertebrates can be found in large numbers. Snails are attracted to the spring-fed fen because of its calcium-rich water; wasps like the overwintering accommodation in the stems of bulrush; and beetles, butterflies, moths, dragonflies and damselflies welcome the diverse food supply for both themselves and their offspring. Birds also value the fen landscape. Resident songbirds like the wren, robin, dunnock, blackbird, song thrush, finches and others are joined by summer visitors like the notorious cuckoo, whitethroat, willow warbler and spotted flycatcher. Open water and reedbeds are occupied by the moorhen, coot, teal, mallard and the little and great crested grebe. In winter, the population of waterbirds grows considerably when the residents are joined by seasonal visitors like the wigeon, shoveler, tufted duck and whooper swan, to name but a few. Fen wetlands are also home to two of Ireland's three amphibian species: the ubiquitous common frog and the rarer smooth newt.

In many places, fens were the precursors to raised bogs. Because dead plant material didn't decompose, the layers of fen peat continued to grow. When the peat layers had eventually reached a thickness that didn't allow the plants growing on the surface to reach the groundwater, or when plant communities were cut off from their mineral-rich freshwater supply in another way, a tipping point was reached. The only food source for the surface plants was now rainwater, which is rather poor in minerals. In Ireland's midlands, this process was supported by the postglacial rebound. The land, finally free from the weight of the ice, lifted itself, and as a side effect the ground-water level dropped. This loss of access to nutritious ground water triggered a radical change in the plant community. Bog mosses, mainly sphagnum species, which are known as the bog builders, moved in. These mosses changed the growing conditions further by lowering the soil and water pH, and in doing so created a more acidic environment. Sphagnum mosses also have the ability to store vast amounts of water, and so they effectively water-logged the surface.

On top of the bog mosses, specialised plants that were able to tolerate the new conditions started to thrive, and so the foundation of a raised bog was laid. Over the following centuries, the mosses and other plants built up layer upon layer of peat, which grew more or less vertically from the lake or pond from which it had originated, creating the characteristic dome shape of a raised bog. Eventually, the single domes merged with neighbouring raised bogs, and so the vast bog areas of the Irish midlands, first and foremost the mighty Bog of Allen, were born.

The combined layers of a raised bog can reach heights of up to twelve metres, but only the top layer shows any signs of life. This top layer, known as acrotelm, is less than fifty centimetres deep and consists of a living growth of sphagnum mosses, the plant communities that thrive on it, and recently deceased plant material. The deeper layers of the raised bog are known as the catotelm. While the upper layer has a high permeability to water, its movement decreases rapidly with

The combined layers of a raised bog can reach heights of up to twelve metres

every layer of peat it passes. The result is a living carpet of vegetation that floats on a watery concoction of dead and compressed plant particles. Around ninety-eight per cent of a raised bog is pure water, and only the remaining two per cent is made of solid material.

While the formation of blanket bogs began in a similar way and at the same time as that of raised bogs, the resulting habitats differ in a number of ways. Not unlike the raised bog, the blanket bog originated in water-filled hallows or shallow lakes and began to form shortly after the last glaciation. For a few millennia, these embryonic blanket bogs only thrived in isolated and confined spots along the coast and in mountainous areas where they were surrounded and even invaded by pine forests. With the aforementioned change in climate, these pockets of blanket bog began to spread. It started with the increased precipitation washing out minerals from the upper soil layers and depositing them further down in the ground, where over time they formed an impenetrable layer known as iron pan. Subsequently, the now nutrient-poor ground became waterlogged, which triggered a change in plant communities and peat growth. This process, however, wasn't entirely straightforward. Slight changes in weather patterns in the early phases of bog formation allowed the forest to recolonise the peatland for limited time periods. The remains of this so-called later pine woodland phase are evident in petrified tree trunks that are a common sight in the blanket bogs, especially on Ireland's west coast.

While raised bogs are dominated by sphagnum mosses, blanket bogs mainly feature grasses and sedges. Sphagnum and other mosses do occur, but in much smaller numbers. Consequently, blanket bogs have a lower water-storage capacity than raised bogs and are dependent on regular precipitation. To keep a blanket bog alive, 1,200 millimetres of rain, drizzle or fog per year, spread out over at least 235 days, are necessary. Another factor that sets raised and blanket bogs apart is their growth. As the name suggests, a blanket bog spreads out over the landscape like a blanket, unlike the raised bog, which takes the vertical approach, fuelled by the fast-growing sphagnum mosses. Blanket bogs, therefore, only reach a depth of two to six metres.

On a global scale, the blanket bog is a relatively rare habitat. Ireland holds eight per cent of the world's blanket bogs, which makes our ones very special. Ireland's blanket bogs escaped the large-scale destruction the raised bogs had to endure but nevertheless were decimated and harvested for fuel or transformed into farmland and commercial forestations. While only an estimated twenty-eight per cent of Ireland's blanket bogs are considered to be fully intact, much of the original flora and fauna manages to hold on in corners of cutover

Opposite top left: Lichen, grasses, sedges and heathers on Atlantic blanket bog.
Opposite top right: Devil's matchstick lichen.
Opposite bottom left: Sphagnum moss on raised bog.
Opposite bottom right: Broad-leaved pondweed and bog pondweed on bog pool.

Above left: Tumduff Mór, County Offaly.
Above right: Pollardstown Fen, County Kildare.

blanket bogs, and in the remote coastal and mountainous areas of the west, large areas have survived almost unscathed.

Roundstone Bog in County Galway is one of the finest blanket bog areas in Ireland. This flat landscape stretches between the lonely mountain of Errisbeg and the granite walls of the Twelve Bens in the south-western corner of Connemara. It is a place that perfectly marries desolation and beauty. Granite boulders stick out of a sea of grasses and sedges, and the whole area is dotted with countless lakes and ponds that reflect the ever-changing sky. All the most interesting bog-dwelling plants can be found here, and some rare oddities in addition. St. Dabeoc's heath has its main distribution on the Iberian Peninsula and makes some sporadic appearances on the coast of France and southern England; its only stronghold in northern Europe is Connemara and the south of County Mayo. How this plant ended up in the west of Ireland is still being debated, but it seems likely that it was introduced by pilgrims or other travellers. The same question of origin hangs over two other heather species. Mackay's heath, which produces no seed and spreads by layering, can only be found in Spain and a few spots on Ireland's west coast. Dorset heath, as the name suggests, is known in Dorset and adjoining counties in England and one other location at Roundstone Bog.

A further kind of peatland which often intermingles with blanket bogs is heath. Its formation and appearance are very similar, but unlike blanket bog, heath retains some degree of drainage. This makes it less waterlogged and results in only a very thin peat layer, which can periodically dry out completely.

While the dominating species groups differ between raised bog, blanket bog and heath, many of the characteristic bog plants can be found in all of them. Heathers, mostly ling (also known as common heather), bell heather and cross-leaved heath, are widespread. In places, patches of fluffy white common cotton grass, hare's tail cotton grass and other cotton grasses stand out, while bog bean, bog rosemary, cranberry and bog asphodel bring other colours to the palette. The latter is one of the most common peatland flowers, but it comes with a sinister reputation. Its Latin name *Narthecium ossifragum* translates to 'little rod of broken bones' and refers to livestock being prone to breaking their ankles after eating the plant. In this case, there is truth in the folklore as bog asphodel contains a chemical substance that inhibits the production of vitamin D, resulting in brittle bones. Bog asphodel is also one of very few plants that has adapted to pollination by rain. Special rain drains guide the floating pollen into contact with the stigma, which, in a wet habitat like an Atlantic blanket bog, considerably increases the chances of successful pollination.

All peatland plants are perfectly adapted to the wet and acidic conditions of these landscapes and have found ingenious ways to make a living in this nutrient-poor environment. The heathers and bog asphodel have formed a partnership with fungi to obtain vital nourishment. The fungal partner, coiled around the roots of the host, provides nitrogen and phosphorus and in return receives carbohydrates it cannot produce itself. The most intriguing solution, however, was found by the carnivorous species. The round-leaved and oblong-leaved sundew, the common and large-flowered butterwort, the lesser and greater bladderwort and the pitcherplant – the last of which was introduced from Canada into County Roscommon in 1906 – have all found ways to catch insects to fulfil their nutritional needs. The sundews modified their either oblong or round leaves to hold long stalks with a sticky drop on the end. Any insect that lands on those stalks gets stuck, the leaf rolls itself up, and the trapped animal is slowly digested. The butterworts catch their prey in a similar way, only here

Peatland plants are perfectly adapted to the wet and acidic conditions

the leaves are spread out in a rosette on the ground and feature a carpet of tiny, sticky hair that holds the victim in place for digestion. The bladderworts do things a bit differently. For one thing, this plant doesn't grow on the bog surface but thrives in the bog pools, with only its small, yellow flowers reaching above the water table. Its underwater stems are fitted with flask-shaped bladders. Once this bladder is touched, a little trap door opens and the unsuspecting meal is sucked into the bladder to be digested.

* * *

The animal life on raised as well as blanket bogs is a mix of a few permanent residents and many visitors. One of the true peatland residents is the red grouse. This reddish-brown coloured bird with characteristically red combs over its eyes can mainly be found on and around heather-covered hummocks. Heather provides food, nesting material and shelter for this quintessential bird of the bogs and heathlands. Due to the destruction of peatlands as well as ongoing hunting, the breeding population of the red grouse is in decline and has halved over the past fifty years. The status of another iconic peatland bird is even grimmer. Over the past few decades, the resident curlew population has been reduced to only three per cent of its original size. The raised bogs of the midlands, together with mountain blanket bogs and the Shannon Callows, were once a stronghold for these birds, but only a few breeding pairs remain today. The main reasons for this are habitat loss and habitat fragmentation, which also drove out another iconic bird. The Greenland white-fronted goose was once a common winter visitor on all peatlands, where it fed on the bulb-like rhizomes of the white beak-sedge. This gregarious bird would breed in summer on the tundra of northern Europe then travel to Ireland and Scotland for the winter months, where it was affectionately known as the 'bog goose'. As intact peatlands disappeared, so did the bog goose. In this case, however, the birds adapted, changed their diet to grass and certain seaweeds, and today the majority of Greenland white-fronted geese spend the winter at the Wexford Slobs, a man-made coastal grassland.

Other resident peatland birds are the snipe, the meadow pipit and the skylark. All three share a similar plumage that in both colour and pattern resembles the wind-blown grasses and sedges of their habitat and so provides perfect camouflage. While these species are not limited to peatlands and are equally common in grasslands, they are an integral part of the bog.

The common frog comes out of hibernation around March, although in mild winters they can be out and about as early as January, which makes both adult and offspring vulnerable to possible cold spells. As soon as the animals are warmed up, mating takes place, and soon after the bog pools are filled with balls of frog spawn. A short time later, tadpoles emerge,

Above left: Large-flowered butterwort.
Above right: Lesser bladderwort.

and over a period of ten weeks they grow into adult frogs. While frogs are commonly associated with water, the adult animals actually only seek out pools and lakes during the mating season and for the rest of the year prefer a more terrestrial life. Therefore, they can be found in a variety of habitats, from peatlands and forests to the stark limestone pavements of the Burren.

The common or viviparous lizard, Ireland's only native reptile, also rises from hibernation around March, and mating takes place soon after. Unlike the common frog, however, the female lizard carries the fertilised eggs inside her body, where they grow into fully developed lizards that hatch after three months. Hibernation as well as their unusual breeding behaviour are an adaptation to the cool climate in Ireland. Further south in the Mediterranean, the viviparous lizard can stay active throughout the winter and lays its eggs

Opposite: Round-leaved sundew.

in a warm and safe spot, as is commonly expected of reptiles. Like the common frog, the viviparous lizard is not bound to the bog and can be found in a variety of other habitats.

While insects and other creepy-crawlies are well represented in fens, they are rather scarce on raised and blanket bogs or heaths. The lack of overwintering opportunities, a limited food supply and fluctuating water levels make these unattractive habitats for most species. Bog pools and lakes provide the best environment for insects, and some are quite happy in this little world. Pond skaters are a common sight on the surface, while the great diving beetle spends most of its life, both as a larva and an adult, under water, using its paddle-shaped legs to glide through its kingdom and feed on tadpoles, up to twenty a day.

Two unique spiders are also at home in the pools. The water spider, or diving bell spider, spends most of its life under the surface. It creates domed webs and fills them with air bubbles — the diving bells that gave the spider its name. The other arachnid is the raft spider, Ireland's largest, which is also known as the Jesus Spider because of its ability to walk on water. It can easily be identified by the broad, white-yellow stripes that run along each side of its body. The raft spider lurks in the mosses at the edge of the pool and dashes across the surface once potential prey is spotted.

The pools also provide the perfect breeding ground for midges and mosquitoes including the infamous biting *Culicoides impunctatus*, which can turn a summer visit to the bog into a bit of a nightmare.

Dragonflies and damselflies can also be found in and around the bog pools. While the nymphs dwell in the bog pools, where they feed on insect larvae and other small animals including tadpoles, the adult dragonflies and damselflies take the hunt to the air. The common darter, brown hawker and large red damsel are the most likely to be encountered and are also widespread outside peatland habitats. More typical peatland dwellers are the ruddy darter and four-spotted chaser as well as the rare variable damselfly, moorland hawker and downy emerald. Watching these elegant insects zigzagging across the bog pools on a warm summer day is one of the great events in nature aviation.

Butterflies and moths are also a regular sight in the peatlands, with some 150 species to be found. Most of them are not limited to peatlands and can be found elsewhere, but one of the bog specialists is the large heath. This butterfly comes in a range of peaty colours and has chosen cotton grasses as its main diet, so it is dependent on intact peatlands. The beautifully chequered marsh fritillary is one of the most endangered butterflies in Europe and feeds only on devil's bit scabious. While this flower is not restricted to peatlands, it prefers a wet environment and has become rare due to the ongoing drainage of farmland.

The wet environment of peatlands should be ideal for molluscs, but surprisingly, these invertebrates are rather rare in this habitat. The only common representative is the black slug, which can mainly be found on blanket bogs. Raised bogs are virtually mollusc free, which, however, doesn't mean there aren't any at all. In 1999, the rare Vertigo geyeri, one of eight whorl snail species known in Ireland, was discovered at Killaun Bog in County Offaly. These tiny snails grow only to a size of two to three millimetres and prefer the outer fringes of the raised bog, where they have access to mineral-rich water. Because these outer fringes were often the first areas of the bog that fell victim to private or industrial turf-cutting, these molluscs lost most of their habitat.

* * *

For the longest time, peatlands were considered useless apart from being a source of free fuel. After the forests had disappeared, burning peat was the only way to heat and cook in many parts of Ireland. Later, peat also became a cheap way to fuel power stations and bring electricity to many parts of the country.

Unfortunately, by the time we had learned about all the real benefits of peatlands, it was almost too late. Bogs are not only a uniquely beautiful landscape with a very special flora and fauna, they also allow us a glance into the past. Under the peat, prehistoric landscapes like the Céide Fields have been preserved, and the layers of peat themselves hold a history of the local flora, climate and atmospheric composition that goes back for thousands of years. Currently, and even more importantly, bogs act as a carbon sink and can store water in times of high precipitation, which can help with flood prevention.

The Irish Peatland Conservation Council and projects like The Living Bog Raised Bog Restoration Project and the Active Blanket Bog Restoration Project have taken on the challenge of protecting the last remaining intact bogs and restoring damaged peatlands. The first results are promising, and it seems that raised as well as blanket bogs recover quickly once the right conditions have been restored. This won't bring back all the lost peat, but it gives some hope for the future. Personally, I will miss the smell of turf fires. But then again, I would rather rest on the squelchy stuff on a hot summer's day than burn it in winter.

Opposite top left: Common cotton grass.
Opposite top right: Hare's tail cotton grass.
Opposite bottom left: Cranberry.
Opposite bottom right: Bog rosemary.

Below left: Common hawker.

Below right: Harvested raised bog, County Offaly.

Bottom left: Meadow pipits on maritime heath.

Bottom right: Heath at Cnoc Mordáin, Connemara, County Galway.

Opposite top left: Irish heath and European gorse.

Opposite top right: Bog asphodel.

Opposite bottom left: Lousewort.

Opposite bottom right: Early marsh orchid.

Above: Heath in the Slieve Mish Mountains, County Kerry.

Opposite: Mountain blanket bog, Wicklow Mountains National Park.

Chapter 7

From Connemara into the Mountains

Many years ago, I had a summer job at a bed & breakfast in Connemara, near the village of Maum. I woke up every morning to a view onto Lugnabrick Mountain, its green slopes sometimes shrouded in mist, sometimes illuminated by the warm morning sun. Once my daily work was done, I spent my time exploring the granite wilderness of the Maumturk Mountains and the Twelve Bens and lost myself in the rolling hills of Joyce Country. I walked bits and pieces of the Western Way, a long-distance walking trail that traverses the Maumturks at Maumean and then follows the western flank of the mountain chain through Inagh Valley, where the Twelve Bens beckon on the other side. From time to time, I left the trail to explore the hillside and enjoy the view from further up. The ever-changing light made shapes and shadows dance along the mountainside, quartzite was glittering under the short-lived spotlight of a sunbeam, and in the distance, a raven croaked.

Opposite: Killary Fjord, Connemara, County Galway.
Below left: The Maumturk Mountains, Connemara, County Galway.
Below right: Winter evening in Connemara, County Galway.

Having fallen under the spell of this bleak but powerful landscape, I returned to Connemara year after year. I experienced an enchanting winter morning at Lough Inagh, the peaks of the Twelve Bens glowing red under the rising sun while the ground was dusted with fresh white snow. I recall mist rising from the peatlands below the Maumturk Mountains, shrouding the lower slopes in a faint haze while the mountaintops were hovering above like a mirage. I remember warm autumn sunshine in the brown fields of bracken that cover the hillsides of the Partry Mountains.

Sometimes when I am looking over the flat landscape of my home in west Clare, I miss the mysterious confinement and embracing shelter of the hills and dream myself to the dark valleys and the towering walls of rock, the lonely corrie lakes and cold mountain streams.

* * *

Below: Connemara coast near Cashel, County Galway.
Opposite: Pyramidal orchid at Ballyconneely Machair, Connemara, County Galway.

Ireland's topography is often compared to a saucer: a flat, low-lying interior surrounded by a series of mountain ranges. Looking at a map of the country, most of Ireland's mountains are indeed neatly arranged along the edges of the landmass. To name but a few, there are the Derryveagh, Sperrin and Bluestack mountains in the north, the Galtee and Comeragh mountains in the south, and the Mournes and Wicklow hills in the east. The west boasts Ireland's highest and arguably most dramatic mountain ranges, among them the Caha and Shehy mountains, the Dingle and Slieve Mish mountains, and the MacGillycuddy's Reeks in the southwest. The northwest is home to the Nephin Beg range and the Dartry Mountains, and right in the heart of the west sit the Mweelrea Mountains, Sheeffry Hills, Partry and Maumturk mountains, and the Twelve Bens. These final ranges complement each other like pieces of a jigsaw and form one massive bulk of mountainous landscape that dominates southwestern Mayo and Connemara.

This is a captivating landscape, rough and utterly desolate but at the same time immensely beautiful. From Galway Bay, the coastline meanders in a northwesterly direction, creating countless inlets, bays, islands and peninsulas along the way. Sheltered sandy beaches alternate with seaweed-covered rocky shores that radiate in muddled but vibrant shades of green, yellow and red. Some of the beaches are made of the finest white sand, while others consist of a much coarser material. Beaches of this kind are known as coral strands, and they are a unique feature of the Connemara coastline. Instead of sand, they are made up of pieces of detached and bleached coralline algae, also known as maërl, and smooth shell fragments, mostly the remains of common limpets, periwinkles, topshells and other maritime snails. The small algae fragments have been broken down and smoothed by wave action, with some resembling tree branches while others appear in a wide variety of geometrical forms. Most have lost their original colour and come in pure white or soft cream tones, while a few have retained some pigmentation and come in variations of pale blue, cyan and grey.

Over time, these fragments get broken down into ever smaller pieces and eventually find their way further up the shore, where in some places they become one of the building blocks of a machair, the rare calcium-rich grassland we came across in an earlier chapter. Ireland holds around fifty machair sites, many of which are located along the Connemara coast and further north in Mayo. Most of the sites are grazed by cattle over the winter, which helps to preserve their wealth of wildflowers. If the machair gets overgrazed, however, by livestock

This is a captivating landscape, rough and utterly desolate but at the same time immensely beautiful

Top left: Roundstone Bog with the Twelve Bens in the distance, Connemara, County Galway.

Top right: The Halfway-House Hawthorn, Roundstone Bog.

Above left: White water lilies, Roundstone Bog.

Above right: Turf stacks, Connemara, County Galway.

Top left: Lough Tay, Wicklow Mountains National Park, County Wicklow.
Top right: Mourne Mountains, County Down.
Above left: Glanmore Lake, Caha Mountains, County Kerry.
Above right: Black Valley, MacGillycuddy's Reeks, County Kerry.

or the ubiquitous rabbits, the plant roots can no longer hold the light soil, which, sooner rather than later, gets blown away and so the grassland disappears.

Further inland, the landscape changes considerably. Where the land hasn't been cultivated to feed livestock or support housing, blanket bog prevails. These peatlands are wide plains of purple moor grass with patches of heathers and cotton grasses, strewn with boulders and dotted with ponds and lakes. In places, turf banks have been cut into the landscape that in summer feature countless stacks and heaps of drying peat briquettes. From the centre of this stark and windswept scene rise the mountains. The Twelve Bens dominate the Connemara skyline, and their rounded domes catch the eye from every direction. To their east, the Maumturk Mountains spread out in a series of peaks, almost seamlessly merging with the bulk of Maumtrasna and the other, gentler hills of the Partry Mountains that overlook Joyce Country. In the north rises the mass of the Mweelrea Mountains and Sheeffry Hills, separated by the eerie Doolough Valley. It's a scene that changes from lovely green slopes to brutal rocks in an instant. There are fields of scree and boulders, sheer crags and gullies, streams and waterfalls, heathy hallows and mossy flushes. It's a landscape of beauty and the beast.

These mountains, just like the Nephin Beg range in County Mayo, the Derryveagh Mountains in County Donegal and the Wicklow Mountains in the east of Ireland, are the product of the Caledonian Orogeny, a mountain-building event that happened some four hundred million years ago. At the time, the mini-continents of Avalonia and Baltica had broken off the supercontinent of Gondwana and were being pushed into Laurentia, another continent sitting in the northern hemisphere. Triggered by the tectonic collision, massive amounts of magma were pushed up from inside the earth and lifted the overlying sedimentary rocks – limestone, sandstone and shale – without breaking the surface. Over time, the magma cooled and turned into granite, an igneous rock and the main building block of the aforementioned mountain chains. Because of the immense heat and pressure during this event, some of the overlaying sedimentary rocks were transformed into metamorphic rocks: limestone became marble, sandstone changed into quartzite, and shale turned into schist and gneiss. These rocks came to sit between the surface rocks and the granite beneath. Over time, the erosive forces of ice and water stripped away the surface, exposed the underlying rocks and started to sculpt the landscape. This process reached its peak during the ice ages, which took hold around two million years ago. During the last phase of the glaciations, an ice sheet more than one kilometre thick and centred over what would become the Twelve Bens covered the biggest parts of the counties Mayo and Galway and also reached south into County Clare. It was this glacier that carved out the valleys and chiselled the mountains into the shapes we see today. The limestones, sandstones and shales disappeared almost completely. The bare peaks of the Twelve Bens and Maumturk Mountains now mostly wear a glittering quartzite cap or display bare granite. In places, veins of marble come to the surface, coloured in the common grey or the unique and rare green of Connemara marble. According to legend, this green marble is a magical form of petrified grass drummed up by the ancient king Rí na Beola. Geologically, however, the green tone would have

The Twelve Bens and Maumturk Mountains wear a glittering quartzite cap

come about when heat and pressure transformed limestone into marble and in the process also created some rare minerals which gave the marble its unique colour.

Another mountain-building event, known as the Variscan Orogeny, began around 250 million years ago and created the mountains of Ireland's southwest. These mountains, which rise along Ireland's south coast and form the backbone of the great peninsulas — Mizen, Sheep's Head, Beara, Iveragh and Dingle — are made of old red sandstone that was laid down around the same time the older mountains further north came into being. Back then, the southern part of Ireland was an alluvial floodplain near the equator, and the regular flooding of this landscape deposited the grains that over time built up to the distinctive coloured sandstone we see today. The rock gets its striking colour, which can range from a deep red to a gentle purple, from oxidised iron particles known as haematite. It is these particles that let peaks like Hungry Hill or the aptly named Purple Mountain glow in warm, rusty tones in the morning and evening sunshine.

As Ireland's south, which was then part of the supercontinent Gondwana, moved northwards, it collided with Laurussia, and over a period of some one hundred million years the forces of colliding tectonic plates folded the sandstone into mountains. The MacGillycuddy's Reeks on the Iveragh Peninsula, the Brandon range and the Slieve Mish Mountains on the Dingle Peninsula, the Caha and Slieve Miskish mountains on the Beara Peninsula, and the Galtee, Knockmealdown and Comeragh ranges further east are all remains of a once massive mountain chain that stretched from what is now North America across Europe and into Asia.

Just like it happened with the older granite mountains, the grinding glacial activity of the ice ages shaped the old red sandstone into its current form. The slowly moving ice sheets cut U-shaped valleys into the rock and scraped out corries, amphitheatre-shaped hollows that are surrounded by sheer rock walls on three sides with an open side facing the valley below. Many of those corries filled with water and exist today as cold and deep corrie lakes, also known as tarns. Wind and rain carved the upper sections of the hills into a mix of sharp peaks and narrow ridges, steep gullies and sheer cliffs, loose scree and boulder fields.

Once these mountains reached heights of over three thousand metres, but today they are rather modest elevations with only a few peaks topping one thousand. The MacGillycuddy's Reeks, or Na Cruacha Dubha, which translates as 'the black stacks', rise just west of Killarney National Park and are the highest mountain group in Ireland. The English name of these mountains replaced the Irish one in the 18th Century and is derived from the MacGillycuddy family, who were among the biggest landowners in the locality at the time. The highest of these conical peaks is Carrauntoohil, at 1,039 metres, followed by Beenkeragh, at 1,008 metres, and Caher, at 1,001 metres. These heights are partly responsible for the local climate and subsequently the famed plant life of the area. While the Gulf Stream keeps temperatures above freezing for most of the year, the mountains catch the moist air coming in from the Atlantic and force it down as rain, drizzle and mist, so it is no surprise that the Killarney region has one of the highest precipitation rates in Ireland. Consequently, the peaks are often hidden in clouds and the forest below drenched in moisture.

While most of Ireland's mountain ranges are the product of the collision of tectonic plates, there are two that tell a different story. The Mourne Mountains in County Down are a compact set of granite peaks, but despite sharing the same kind of rock, they are much younger than their neighbours in counties Donegal, Mayo and Galway. The Mournes rose up only around sixty million years ago, when Ireland's northeast was shaken by an era of high volcanic activity that also created the Antrim Plateau and the Giant's Causeway. During this event, a batholith, or mound of magma, rose under the future Mournes and created a massive dome. This dome was then carved into the distinct succession of round peaks and U-shaped valleys by the glaciers of the ice age. The Mournes are one of the most recognisable

Below: Raven.

Opposite: Mossy saxifrage.

mountain ranges in the country, and their characteristic outline commands the landscape for many kilometres around. Seen from a distance, they are often covered in a blue haze and seem to rise straight from the jigsaw of fields and hedgerows. Once you get closer, you can make out the lower slopes, covered in a thin layer of blanket bog and heath, and the bare, grey granite peaks, smoothed by millennia of ice, snow, rain and mist. Another feature that will catch your eye is the Mourne Wall, an enclosure built between 1904 and 1922, which traverses up and down the mountains, scaling fifteen peaks in the process. The proximity of the Mournes to the city of Belfast made them an ideal source of water for the expanding city. A reservoir was built, and the wall was erected around the catchment area to keep sheep and other unwanted visitors away from the water source. Over time, the wall has not only become a landmark but also a handy navigation guide for hill walkers.

On the other side of the country, some 150 kilometres as the crow flies, resides an unusual and unique – for Ireland, at least – set of elevations. Ben Bulben and the surrounding hills that line the border between counties Sligo and Leitrim, collectively known as the Dartry Mountains, are Ireland's only set of table mountains. The rocks that make up these immediately recognisable hills are limestone and shale that formed some 320 million years ago. Before the repeated glaciations during the ice ages, the area was a long, continuous ridge, but the movement of the ice cut deep valleys into the rock and left the remaining parts of the ridge as flat-topped mountains. This unique topography allowed a special set of plants to survive on top of the plateau, which is several degrees cooler than the plains and valleys below. Pollen analysis has shown that parts of the tundra flora that was common at the end of the last glaciation around ten thousand years ago has continued to thrive on the summit of Ben Bulben. This flora includes one plant that can't be found anywhere else in Ireland or neighbouring Britain. The fringed sandwort is a typical wildflower of the Arctic and high mountains of central Europe, which forms flat tussocks on the rocky ground and displays small, narrow leaves and delicate white flowers. Its continued presence on Ben Bulben is not easily explainable to begin with, considering Ben Bulben is neither very high nor anywhere near the Arctic, but a study has also shown that this cold-loving plant has not only survived since the ice age but also *through*

Tundra flora from the end of the last glaciation around ten thousand years ago has continued to thrive on the summit of Ben Bulben

 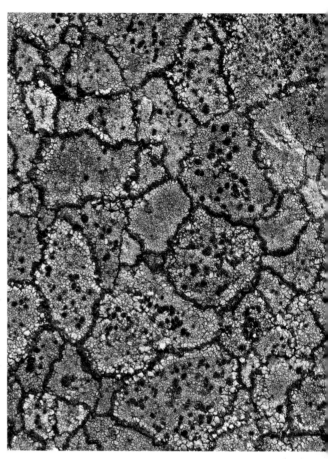

Above left and right: Lichen landscapes.
Opposite: Ben Bulben, County Sligo.

it, which makes the fringed sandwort even more mysterious. Scientists now believe that the plant has been growing on Ben Bulben continuously for some 150,000 years.

Dedicated alpine plants are generally rare in Ireland. Because of our mountains' modest heights, blanket bog extends to all but the highest peaks, bringing with it its characteristic flora. Only where the peat didn't take hold, on the bare rocks and scree fields, can something of a mountain flora be found. Mosses, liverworts and lichen are most common. Connemara, in particular, has gained some fame among lichenologists. The variety of rock types and different habitats support close to seven hundred species, which grow on the exposed rock in the mountains and the coast, the boulders that are strewn across the boglands, and also on the bark of trees. These lichen often form intricate, map-like structures and come in striking colours of orange and yellow. It is theorised that these colours are directed at birds, indicating a safe place to perch. The birds, in return, leave droppings rich in nitrogen that is absorbed by the lichen.

Some rare mosses are also at home in the mountains. *Orthothecium intricatum* was discovered

in 1968 and is endemic to Ireland and Britain, while *Oxystegus hibernicus* can only be found in western Ireland and Scotland. Equally rare is the alpine clubmoss, found in just a handful of locations around Ireland. Clubmosses, despite their name, are not actually mosses but closely related to ferns, and because of their appearance, they are also known as ground pines. Real ferns, meanwhile, seek out crevices in the rock and other somewhat sheltered areas. In addition to some common species like the brittle bladder fern and broad buckler fern, the upper slopes are home to two rare and truly Arctic-alpine species, the holly fern and the green spleenwort, which are both limited to Ireland's northwest in their range.

Typical but overall rare wildflowers of these higher elevations are the mountain sorrel, mountain avens, alpine saw-wort, alpine meadow-rue, and yellow, purple and other saxifrages. The saxifrages are probably the most widespread high-mountain dwellers; their genus name, *Saxifraga*, literally means 'rock-breaker'. Roseroot, another plant of the high mountains, got its name from the fact that its crushed root emits a smell resembling roses. It has been used in Scandinavian countries for centuries to treat anxiety, fatigue and depression. Researchers have found at least six categories of antioxidative and stress-mediating compounds in the root of the plant. In other words, the Vikings knew what they were doing. Thrift and sea plantain, which are traditionally associated with the coast, are also common in the mountains, as are sheep's bit and sweet alison. Also a bit out of place is the lesser twayblade, the only orchid that can grow in heights up to 490 metres. St. Patrick's cabbage, a member of the Lusitanian flora, is also not the typical mountain plant but rather common in the Connemara uplands.

Animal life is also rather scarce in the hills. The largest creatures there are the native red deer and the introduced sika deer. In some areas, feral goats roam lower slopes of the mountains. The majority of the remaining mountain fauna are birds. The raven and peregrine are the most common species, with the raven associated with remote upland areas. The hoarse croak of this beautiful black bird, circling around lonely peaks, is the epitome of a wild mountain scene. The peregrine, on the other hand, is less obvious and usually silent and can mostly be seen hovering close to the ground looking for its next meal. This bird of prey came close to extinction in the 1960s, when the use of pesticides became widespread. The poison didn't affect the birds directly but travelled through the food chain and accumulated in the peregrine's body, with the

Animal life is also rather scarce in the hills. The largest creatures there are the native red deer and the introduced sika deer

result that the female birds' eggshells became very thin and brittle. What followed was a quick and steep drop in the number of young peregrines. After the pesticides responsible were banned, the peregrine made a remarkable comeback and is one of the most common birds of prey today.

The wintering snow bunting — a small bird with a short but distinctive yellow beak with a black tip — can sometimes be seen on exposed mountaintops or feeding in groups further down. The ring ouzel — a close relative of the blackbird and similar in size and appearance with the addition of a white crescent on the breast — was once a common summer visitor to the remote scree slopes where it reared its young. Today, the bird can only be found in a few places in Kerry and Donegal.

Away from the rocky cliffs and slopes, the mountains are covered in a patchwork of blanket bog and heath, which are dominated by grasses, rushes and heathers. Small mountain streams have carved their beds into the peat and the underlying rock, and after heavy rain these waterways grow into raging torrents. This is the domain of the dipper, which resembles the ring ouzel but is slightly smaller and displays a larger and very prominent white bib. The dipper living in Ireland is a unique subspecies, which shows a rusty brown band just under its bib. This bird prefers fast-flowing streams, where it feeds on the larvae of caddies and mayflies, and also builds its nest under small waterfalls or overhangs.

I had my first encounter with a dipper many years ago at the Connemara National Park. It was just outside the visitor centre, which is also the starting point for one of the most beautiful (as well as easiest) mountain hikes in the country. Diamond Hill, which belongs to the Twelve Bens range, rises only to a height of 442 metres but offers the most astonishing views: to the north and west lies the intricate Connemara coastline and its islands, to the south and east rise the glittering Twelve Bens and the rolling hills of Joyce Country.

Ireland's mountains lack the height of their relatives on the European mainland, but they very much make up for that with character and drama. Mist swirling around the peaks of the Twelve Bens, the rising sun creeping up the slopes of the Maumturk Mountains, or a dark wall of clouds hanging between the Mweelrea Mountains and the Sheeffry Hills is as beautiful as the natural world can get.

Above: Partry Mountains, County Mayo.

Above: Diamond Hill, Connemara National Park, County Galway.

Chapter 8

Ocean Travellers and Coastal Dwellers

They seem to appear out of nowhere, filling the sheer rock shelves and cliff tops of Ireland's coast until there is no space left and the rock itself transforms into a mass of shifting black and white bodies. These are the ocean travellers, the ghosts and conquerors of the sea; the guillemots, razorbills, fulmars, kittiwakes, Manx shearwaters and puffins.

It is an event that repeats itself every spring. The feathered globetrotters return to reclaim their nesting spots, mate and rear another generation, before secretively disappearing again in late August. For centuries, the life of these birds was a mystery, and they found their way into myth and legend and ancient texts like Homer's *Odyssey* and 'The Seafarer'. Today we know that these mysterious summer visitors spend the winter on the open ocean, to rest and feed after a tough summer of parenting. During this time, they cover vast distances; some visit rich feeding grounds on the other side of the Atlantic or veer north towards the Arctic circle, while others travel south to the Mediterranean or even further to spend the winter months off the African coast.

But wherever they venture, from late January onwards the birds start to return to Ireland. First only single birds arrive for brief visits, but later they come in groups and stay for longer periods between their ongoing fishing trips until eventually the colonies are complete in late April. For us coastal land dwellers, this is a sign that spring has arrived. Seeing the comings and goings of a busy bird metropolis, the air filled with bird talk and the aroma of guano, we close our eyes and take in the sound and smell of the coming summer.

* * *

One of the first to arrive back on terra firma every year is the fulmar. Every year, I look forward to my first encounter — it is like welcoming back an old friend who has been away on a long journey. The bird swiftly and elegantly glides past me, an inquisitive eye checking me out, before a minute movement of its wing and tail feathers sends the traveller into a dive and out of my sight.

Opposite: Gannet.

Opposite: Winter storm.

The fulmar – or to be precise, the northern fulmar, scientifically known as *Fulmarus glacialis* – is one of the great ocean travellers and related to the petrels, shearwaters and albatrosses. Because of its plumage, a white head and a white body with grey upperparts, the fulmar is often mistaken for a gull. A closer look, however, reveals a stockier build and a curious-looking beak. Unlike that of the gull, the fulmar's beak consists of several horny plates with tube-shaped nostrils, known as naricorns, sitting on top. These nostrils are used to excrete excess salt from the body, and they also give the bird an acute sense of smell – a very helpful ability when it comes to locating food on the vast expanse of the open ocean.

On land, the fulmar appears like the clumsy character out of an old slapstick movie. It drags its body along, apparently not quite sure how to use its legs, always close to toppling over, and resting after every other step. Once the legs have parted with solid ground, however, the fulmar's demeanour changes completely. Confidence returns to its eyes, and its body transforms into a sophisticated flying machine. The fulmar is built for living its life on the wing. It is a master of the wind, and the stronger it blows, the more the fulmar is in its element.

Up until the early years of the 20th Century, the northern fulmar was a relatively rare bird, its distribution limited to a few colonies in remote parts of the north Atlantic, including the islands of St. Kilda, off the coast of Scotland, and Iceland. For the islanders, the fulmar was an important part of their livelihood, providing not only meat but also feathers for bedding and oil for lamps and general ailments. On St. Kilda, fulmar's oil was used to cure anything from rheumatism to toothache. It was on these islands that the fulmar got its name, *fúll már*, the old Norse for 'foul gull'. This name refers to the bird's rather unappetising defence tactic of spraying stomach oil onto potential predators. This smelly and sticky oil is stored in a special section of the fulmar's stomach called the proventriculus, which sits between the oesophagus (the food pipe) and the gizzard (the stomach proper). During long flights, the fulmar also uses the oil as an emergency ration, and during the breeding season, to feed its offspring if prey gets scarce.

In the first decades of the 20th Century, the fulmar begun to spread quickly across the north Atlantic. The reason for this population explosion was never fully understood. It seems likely, however, that at the time a change in the islanders' diet – and therefore a reduction in the number of harvested birds – on one side and increased commercial fishing activity and subsequent discards on the other all might have favoured the fulmar.

The fulmar is built for living its life on the wing. It is a master of the wind

Above: Fulmars.

Ireland welcomed its first breeding pair into County Mayo in 1911, followed by another pair that settled in County Donegal a year later, and from there the fulmar extended its range all around the Irish coast. The following decades saw fulmar numbers across the north Atlantic soar to over four million pairs. In the latter half of the 20th Century, these numbers started to decline again, and some estimates suggest that up to forty per cent of the north Atlantic population has been lost. In Ireland, however, fulmar numbers are somewhat stable. The latest breeding bird count in the year 2000 found close to 33,000 pairs, which confirmed a stable and even slightly growing population compared to previous counts in 1970 and the late 1980s.

The main reason for the overall decline is thought to be plastic pollution. Studies in the North Sea have shown that over ninety-five per cent of all dead fulmars had plastic particles in their stomachs. On average, the birds had ingested forty-four pieces of plastic, but one bird was found with a total of 1,603 pieces and another one with a staggering 20.6g of plastic in its stomach – equivalent to around 2kg in a human-sized stomach. It is likely that these birds picked up plastic while skimming the water surface for plankton, or they could have mistaken floating plastic pieces for fish. Fulmars can live up to forty years, but with their crop (a muscular pouch where food is stored before it moves on through the digestive system) and stomach filled with plastic, they can neither feed their offspring nor themselves and would starve to death well before their time.

The fulmar's diet is rather varied and includes plankton, fish, squid, crustaceans and carrion. On their foraging journeys, these birds can cover vast distances – they have been recorded travelling some 2,500 kilometres away from their breeding colonies. One individual managed to fly 1,600 kilometres in only fifty-five hours to visit particularly rich feeding grounds along the mid-Atlantic ridge. During the winter months, fulmars travel even further. Birds belonging to breeding colonies in Ireland and Britain have been recorded on the coasts of Greenland, Newfoundland, Nova Scotia, the Netherlands, Denmark, Iceland and the Faroes. Intensive studies of the fulmar's travel pattern have also revealed that these birds have very individual and different strategies. Some birds stay relatively close to their breeding sites all year round, while others travel far and wide in search for the richest feeding grounds. Most birds stay true to their personal routes year after year.

No matter their travel habits, all fulmars return to the coast to breed. I am lucky to have a small colony of these birds on my doorstep in west Clare, and 'my' fulmars return from their winter travels every year in late January to get reacquainted with their mates and nesting sites. It's a first sign of spring, fulmars soaring along the cliff edge and resting in pairs on the narrow ledges while exchanging gentle caresses and performing their cackling dance. Unlike the cheeky herring gull, the flustered guillemot or the nonchalant puffin, the fulmar always brings an understated curiosity to a chance meeting. Its dark eyes stay fixed on me while the bird soars past at eye level and turns to do another flyby, then turns again for another look. No other seabird I have ever met does this. In March and April, the fulmars usually disappear again for a while. First the female goes on an extended foraging trip, building up reserves for the breeding season. Then the male does the same but usually stays closer to the coast, most likely to keep an eye on — and if necessary, defend — the breeding site.

Its dark eyes stay fixed on me while the bird soars past at eye level. No other seabird I have ever met does this

The breeding season starts properly in mid to late May, when the female lays a single egg. After a short incubation by the female, the male takes over, and in the following weeks the parents alternate their duties in shifts of around nine days during incubation and five days after the chick has hatched. The chick exits its shell around late June or early July as a ball of greyish fluff. The young fulmar grows quickly on the rich diet its parents provide and turns into a big lump which doesn't resemble the sleek adults in the slightest. This overfeeding is deliberate, giving the young bird the necessary reserves for starting its own life in a rough environment. After a few weeks, the young fulmar starts to lose its downy gown, and at this stage, both parents fly away from the nesting site to bring back food, leaving their hungry offspring unattended. By then, the youngster is well able to defend itself against any predator, and some even take their first wobbly steps away from the nest. The breeding season ends in late August when first the parents, and a few days later, the fledgeling leave the cliffs they have spent the summer on.

At a glimpse, the kittiwake appears similar to the fulmar and often their colonies can be found in close proximity. A closer look, however, reveals a more delicate-looking bird, with dark eyes surrounded by a snow-white head and chest plumage, a yellow bill and a distinctive red mouth and throat. This medium-sized gull, the only one in Ireland to nest exclusively on sheer cliffs, is named after its piercing call — 'kittee-wa-aaake' — and has been roaming the seas for millennia. Descriptions of birds resembling kittiwakes make an appearance in stories as old as Homer's *Odyssey*, written down some three thousand years ago and compiled from even older tales.

Today, the kittiwake is one of the world's most threatened seabirds. It is estimated that there are eighteen million birds living in the northern hemisphere, and while this sounds like a decent number, kittiwake populations have declined by forty per cent since 1970.

Opposite top and opposite bottom right: Kittiwakes.
Opposite bottom left: Juvenile kittiwake.

Top left: Breeding kittiwakes and guillemots.

Top right and above right: Guillemots.

Above left: 'Bridled' common guillemot, a genetic polymorphism that manifests itself in white circles around its eyes that stretch back as a thin white line.

Looking at local numbers, the downward trend becomes even more obvious. On the Orkney and Shetland islands, numbers have plummeted by eighty-seven per cent since 2000, and on St. Kilda, a staggering ninety-six per cent of these birds have disappeared. The Irish numbers are not much better. The colony on Ireland's Eye, off the Dublin coast, has lost fifty per cent of its breeding pairs, leaving only around four hundred pairs today. The largest colony in Ireland can be found at the Cliffs of Moher, with seven thousand pairs, but there, as well, the numbers are dwindling.

Overfishing and climate change, and the resulting lack of food, are to blame. Unlike the diving auks – the puffin, razorbill and guillemot – kittiwakes are surface feeders. They skim their prey in mid-flight from the water or plunge to grab a meal from just below the surface. During the winter, when fending for themselves only, kittiwakes can cover vast distances to find food. In summer, when they have not only themselves but also hungry chicks to feed, their travelling time is limited, and they tend not to stray more than fifty kilometres from the coast. The distribution of the kittiwake's prey, mainly sand lance, capelin and other small fish, depends on ocean temperature and ocean currents. Rising average temperatures associated with climate change can upset the balance that times the arrival of fish in coastal waters and the breeding season of the birds. Throw industrial-style fishing into the mix, and you end up with an area of empty ocean where plenty of fish are supposed to be. When this happens, the birds are in trouble.

Statistically, kittiwakes need to fledge on average 0.8 young each year to maintain a stable population, and in many colonies, this threshold is not being reached anymore. Kittiwakes lay between one and three eggs, with two being the norm for most birds. To successfully rear the chicks emerging from these eggs, kittiwakes need a stable food supply within a certain area of their nesting site. The birds are formidable and clever hunters who know where and when fish are supposed to appear. They consider ocean currents and tidal conditions, and if necessary, they will substitute fish with marine worms, molluscs and crustaceans. Because of this, kittiwakes can still manage even when fish numbers plummet as low as seventy per cent below normal. More and more often, however, the required food just isn't available, which leaves even the most experienced birds hungry. Eventually, the parents have no other option than to leave their nest and chicks behind to travel further onto the open ocean to find food.

Kittiwakes return from their winter break to their established breeding sites, which they often share with guillemots, in late March and early April. Here, they build a cone-shaped nest, which traditionally consists of seaweed and grass, but

The birds are formidable and clever hunters who know where and when fish are supposed to appear

more recently plastic and pieces of fishing rope or net are incorporated as well. When nesting material is difficult to come by, kittiwakes have no problem stealing bits and pieces from neighbouring nests, which often results in noisy scuffles. These aggressive and mostly short-lived confrontations are not restricted to the nesting site itself – I have seen opponents falling off the cliff edge in the heat of battle, rearranging themselves and then continuing the dispute on the water, creating a confusing but strangely elegant water ballet. The kittiwakes' noisiness is not limited to their disputes; their colonies are continuous walls of sound. The birds shout when fighting with their neighbours, they scream to greet their partner coming back from a fishing trip, they yell when another bird flies by – or for no apparent reason at all.

After courtship, the female kittiwake produces brownish to greenish speckled eggs, and after a few weeks of incubation, white chicks emerge. While the kittiwakes' behaviour is in parts comparable to that of other gulls, the newly hatched chicks show the alternative evolutionary road that is engrained in this species' DNA. Most gulls are ground breeders, which means their chicks can leave the nest and wander around, even if it is only for a few centimetres. For kittiwake chicks, nesting on the cliffs, this would mean certain death, which is why they stay put in the nest. Ethologists Esther and Mike Cullen wanted to know if this behaviour is really part of their DNA or dependant on the actual location of the nest, so they conducted a cruel but revealing experiment in the 1950s. They replaced kittiwake eggs with ones from ground-nesting black-backed and herring gulls. Kittiwakes can't differentiate between their own and other eggs and so continued breeding until the chicks hatched. All the black-backed and herring gull chicks left the nest as soon as they could walk and plummeted to their death.

Zoologist Heather McLannahan took this experiment even further in the laboratory. She built an artificial cliff, where a transparent Perspex glass extended from the cliff edge. Then she hatched kittiwake chicks, blindfolded them, and put them at the very edge of the fake cliff. The reactions of the young birds once the blindfold was removed were striking. After looking around, each chick would crouch once, start trembling all over its body, try to hook its claws into the underground and then turn and crawl away from the apparent abyss. Once a few centimetres away from the edge, the bird would visibly relax. Herring gull chicks confronted with the same setup would happily crawl onto the plexiglass.

While kittiwakes' nesting behaviour is precarious enough, the guillemot takes this life on the edge another step further. A guillemot colony shares many similarities with a New York subway at rush hour: the birds are tightly packed into a limited space, noise levels are high, and there is a constant coming and going. The biggest of these colonies can be found at the Cliffs of Moher. According to the Seabird 2000 census, it would have been home to nineteen thousand birds; Loop Head in Clare was found to have over five thousand; Horn Head in County Donegal, 6,500.

Above: Razorbills.

Opposite top left: Black guillemots.
Opposite top right: Puffin.
Opposite bottom: Gannet colony on the Great Saltee, County Wexford.

Consequently, the guillemot has the smallest breeding territory of any bird, around 18cm across, or as far as the bird can stretch its neck to tussle with its neighbours. Under these conditions, it would be futile to build a nest, so guillemots lay their single egg directly onto the rock surface. This speckled egg is rather narrow, pointed and somewhat pear-shaped, and one theory suggests that this shape makes it less likely for the egg to accidently roll off the ledge. The egg is laid in early May, and the guillemot parents share incubation duties so both can go on fishing trips to keep their strength up.

When observing a guillemot colony, it is striking that there is always a crowd of birds on the water as well as on the cliffs. This aggregation on the water is the guillemots' information hub. The birds that are about to go on a fishing trip sit and watch returning birds, taking the direction from which these birds return as an indicator of where good hunting grounds may be on the day. How much the squabbling and shouting plays a role in this information exchange, we don't know. Like all birds, guillemots plan their fishing trips well; they know from experience where fish can be found at any given time but also take on information from their fellow birds and adjust their outings accordingly. They know exactly what they are doing and leave nothing to chance.

Guillemots feed on small fish and crustaceans, and like all auks, they are excellent divers. They dive from a floating position on the surface and use their short wings to propel them downwards. These dives can last up to four minutes and go to a depth of almost two hundred metres. Unlike those of land birds, guillemots' bones are solid, which provides them with ballast and protects them from the rising water pressure further down. Their eyes have also adapted to their hunting style. They fill almost the entire skull, with only the large pupil visible, which gives the birds extremely good sight even in the semidarkness deeper down the water column.

The guillemot chick hatches in July as a brownish ball of fluff and is looked after by both parents, who share fishing and caring duties. Only three weeks after hatching, the youngsters have developed into a miniature version of their parents and are ready to leave the nesting site. They do this by simply jumping off the ledge into the sea, a mysteriously orchestrated event that sees all fledgelings leave the nesting site at the same time and mostly under cover of night. When in the water, the father will take over parenting duties permanently for another ten to twelve weeks, until the chick has learned to fish. During this time, father and offspring slowly travel away from the coast onto the open sea.

The youngsters stay on the ocean for their first three to four years, and only then do they start to visit breeding colonies, watching and learning from the older birds before they are

Opposite: Breeding gannet.

ready to breed themselves, which happens after around six years. Once settled, they will return to the same breeding spot for the rest of their lives. This could be several decades; the oldest known guillemot reached a proud forty-three years old.

Like other species, guillemots are affected by decreasing fish numbers. Ever more often, both parents have to go fishing at the same time, leaving the chick unprotected, and even then they are often not able to catch enough food to rear the young bird. The outcome is a higher mortality rate among the young guillemots. On Loop Head, numbers have toppled from five thousand in the year 2000 to four thousand in 2010 and just under three thousand in 2021.

The razorbill is a close relative of the guillemot – both are members of the auk family. They are similar in appearance, but the razorbill is slightly smaller and plumper, with a characteristically large black bill featuring white lines running from the eyes to the tip of the bill. A guillemot's bill, in contrast, is slender and pure black. Both species often share the same breeding space, but while the guillemots huddle together in the open, razorbills seek out cracks, crevices and ravines in the cliff face to lay their eggs. Razorbills are also known to change their partners from time to time, unlike guillemots, who mate for life.

Otherwise, the two species are quite alike: like the guillemot, the razorbill lays one egg; the chick hatches after around five weeks of incubation and is fed on a similar diet; and both species dive to catch their prey. After fledging, like the guillemot chicks, the razorbills jump off the cliff, and the chick stays with the male parent for a number of weeks to learn how to fish. And like those of the guillemot, razorbill numbers are also dwindling fast.

Ireland's best known and most beloved member of the auk family, the puffin – known as *fuipín* or 'chicken of the sea' on the Great Blasket – faces the same predicament, and the reason for its decline is also a lack of food. Puffins return to their breeding sites in late April. Before that, however, they undergo a wonderous transformation. The winter puffin is very different to its summer alter ego; its beak, legs and face show a muddled grey colour, while the body is clothed in dull white and black. Once they days get longer, the birds begin to change. The pineal gland in the hypothalamus and other light-sensitive cells in the brain register the lengthening of days, which triggers the release of hormones. These hormones set in motion a series of events. First the birds feel the urge to feed more. Then there is a complete change of feathers and the production of pigments that find their way into the beak and legs. The distinctive colouring of the beak comes from carotenoids in fish, so a colourful beak indicates a strong, healthy and well-fed individual. Finally, in male birds, the testicles grow from the size of poppy seeds to the size of broad beans.

After the birds have returned to their breeding sites, clothed in their full summer colours, they reacquaint themselves with their partners. Puffins don't necessarily mate for life, but the divorce rate is very low. After mating, they prepare the nest, which is an underground burrow on a grassy clifftop. Where possible, puffins use abandoned rabbit homes; if there are none, they dig one themselves, and where digging is not possible, they find alternatives under boulders or in cracks in the rock. Recent observations at breeding sites in Britain and Iceland revealed that puffins are not only pretty but also cleverer than we thought. Several individuals were caught on camera using small sticks to scratch their backs and chests, most likely to remove ticks and other parasites.

The female lays one egg, which weighs around twenty per cent of her body weight. Like its relatives, the puffin feeds on small fish, caught on deep dives. On average, puffins dive to a depth of fifteen metres but can go down to more than two hundred metres and stay there for up to two minutes. To feed itself and its offspring, a puffin has to go on around eight food runs every day, and during each run, the bird performs around one hundred dives. Some 450 sandeel are needed each day to make sure the chick gets from its birth weight of 60g up to 350g, the weight needed to fledge and leave the nest.

The youngsters that make it take off to sea, where, similar to the other auks, they spend the first five years of their life exploring and finding their own routes for their winter journeys. Research has shown that each puffin has its own individual winter schedule that they meticulously follow every year. After their wandering years, most puffins return to their place of birth to find a partner and breed.

The Manx shearwater is the only other seabird that breeds underground, and it is therefore no surprise that they can often be found in the same spot as the puffin. Shearwaters are relatives of the fulmar and are most elusive birds. They spend the winter far away in the southern Atlantic and return to their breeding sites on the Blasket Islands in Kerry, the Saltees in Wexford and Copeland Island in Down in springtime. There, they spend the days underground in their burrows and only leave under cover of darkness to go fishing, a unique behaviour that raises the ghost comparison to a whole new level.

The gannet is one of Ireland's most elegant birds. Even from a distance, gannets have an air of aristocracy about them, gliding on stiff wings over the water, unfazed by the world around. Up close, this bird retains its distinguished appearance: piercing blue eyes, a buff cap that fades into the pure white of the body, a long bill of light cerulean colour decorated with black

After the birds have returned to their breeding sites, clothed in their full summer colours, they reacquaint themselves with their partners

Top left: Herring gull fight.

Top right: Herring gull raiding a nest.

Above left: Herring gull with spiny starfish.

Above right: Greater black-backed gull.

Top left: Cormorant.
Top right: Cormorant chicks.
Above left: Juvenile chough.
Above right: Chough.

lines, and an unchanging, stern expression make this a being that inspires awe and respect.

Unlike most other summer visitors, the gannet doesn't set foot on the mainland; instead, these birds form large breeding colonies on remote offshore islands. The largest can be found on Little Skellig off the coast of Kerry, where around 26,000 breeding pairs and many thousands of non-breeding individuals were recorded in the Seabird 2000 survey. Further colonies are located on Bull Rock in Cork, the Great Saltee in Wexford and Ireland's Eye in Dublin.

Each of these colonies has its own fishing grounds, which are respected by the other colonies. The Little Skellig birds, for example, always head north and can be observed fishing around the mouth of the Shannon, while their next-door neighbours on Bull Rock exclusively travel south on their fishing trips. Gannets mainly feed on mackerel and herring, which they attack in a unique and impressive fashion. Their keen eyesight allows them to circle high above the water, scanning the ocean for potential prey. Once a target is spotted, the bird turns itself into a sleek missile and plunge-dives from a height of around twenty to forty metres and at a speed of up to one hundred kilometres per hour. Anybody who ever jumped into water from a height knows that the impact can be rather painful; the gannet dampens this with the help of interconnected air sacs that are strategically placed around the body. Once the bird is in the water, the air stored in the sacs is returned to the bird's lungs, where it is used as an oxygen resource during the dive. The initial impact speed takes the gannet to a depth of around ten metres; from there, the bird can follow its prey down to twenty-five metres.

A gannet colony, also known as a gannetry, is a sight not easy to forget and reminiscent of an oversized chess board. Each square houses a round nest in its very centre, made of dirt, debris and seaweed as well as other plant material. The size of the squares is determined by the reach of the occupants' beak thrust. A gannetry is not a peaceful place. From the arrival of the first birds in spring and the first fights over nesting sites to the fledging of the young, there is constant grappling and brawling going on. Fights for nesting sites, which are always carried out by members of the same sex, can last for hours, and any bird that enters a nesting site other than its own is made painfully aware of its mistake. Wandering chicks are likely to be killed by thrusting beaks.

Each year, the older birds are the first to arrive at the colony and reclaim the prime spots at the centre. Younger birds arrive a bit later and share the remaining space with non-breeding juveniles around the edges of the colony. Once all spots are filled, the couples get reacquainted by ritualistic play fight, which is beautiful to watch. There is beak fencing, breast touching and sky pointing, and all the while the dignified and elegant behaviour of the birds takes centre stage.

Gannets lay only one egg but can produce a second if the first gets lost early in the breeding season. The newly hatched chicks are initially featherless and develop a fluffy coat during the first week. They are ready to leave the nest after around thirteen weeks, and this coming of age is a 'do or do not, there is no try' situation. While other young birds start their journey into adulthood by drifting on the water or practising their flight on land, the young gannet just takes off and hopes for the best.

* * *

Not all the feathered creatures we like to call seabirds consider the open ocean their home and only return to land to breed. Some have adapted to the allegedly more convenient terrestrial life and settled at the coast. Gulls are by far the biggest group of coastal dwellers, and some have even moved further inland. The common gull or black-headed gull can be seen at lakes or along rivers, and others have even adapted to city life.

Among those are the most common — and least loved — species: the herring gull and the greater and lesser black-backed gull. While these species were once truly coastal, and in places they still are, they have adapted to a world shaped by humans and are now true birds of the Anthropocene, in many ways not so different from ourselves. These gulls are clever opportunists and have learned that we are the source of an easy meal. This starts with the fish leftovers on trawlers and in harbours and goes as far as individuals directly targeting people, stealing food from their hands. While this might seem cunning and funny, it also speaks of a high intelligence. A group of scientists spent a sunny summer afternoon in a park in Paris, where people were feeding bread to the local mallard population. All of a sudden, a herring gull rushed in, grabbed a piece of bread, then retreated to the centre of the pool, where it crushed the bread and spread the pieces onto the water surface. Not long after, an orange shape ascended from the depths of the pond; the gull struck and had herself a nice goldfish meal. After that, she went back to the mallards for more bread.

Gulls and many other birds are keen observers and able to learn and adapt their behaviour to make their lives easier and better. It is not by chance that gulls started nesting on buildings and hanging around on promenades — for them, it's like living in a penthouse right beside a delicatessen.

Two other well-known coastal dwellers are the shag and cormorant. Their long, snake-like necks, hooked bills and shimmering blackish-bluish plumage sets them apart from all the other seabirds. Today shags and cormorants are different, but they are the shadow

Not all the feathered creatures we like to call seabirds consider the open ocean their home

of the origin of all seabirds. Their ancestors go back in a direct line to the time of the dinosaurs, and the fact that cormorants are not only coastal dwellers but can also be found inland shows that they have a firm foot on land as well as in the sea. While the new species of seabird that emerged over the millennia perfected living on and in the ocean, the cormorant and shag held on to their traditional lifestyles and therefore barely changed physically. They are built for a life near water, be it the ocean, a river or lake, but need firm land upon which to rest and dry themselves after fishing. They are too heavy to fly long distances and can only dive in shallow water, where they propel themselves to the bottom to hunt for fish. Their breeding behaviour is also very different to those of the younger species of seabirds. They don't mate for life, and they build big nests and lay several eggs, just like their ancestors.

There are other coastal dwellers, but none of them are considered seabirds. After the disappearance of the forests, the raven adapted to a life on the coast. The largest member of the corvid family builds impressive nests on the rock ledges of cliffs and only recently extended its range into a similar habitat in the mountains. A close relative of the raven, the chough, with its distinctive red beak and legs, forages in grasslands on clifftops and nests in cracks and crevices of the rock face. The rock dove, an ancestor of the domestic and feral pigeon, leads a similar lifestyle. The rock pipit is another common coastal bird, and in places starlings have chosen to breed on the cliffs ... and who could blame them? Ireland's coast is a wonderful place to live.

Above: Raven foraging.

Opposite: Raven at its nest at the cliffs.

Opposite top: Rock pigeons.

Oppoiste bottom: Juvenile herring gull.

Above: Rock pipit.

Postscript

Autumn sunshine is flowing into my office, birds are singing in the hedgerow bordering my garden and the Atlantic Ocean lies calm and wide outside my window. The world looks perfect and beautiful yet I am trying in vain to write a positive ending to this book. After a summer that brought – yet again – unprecedented heatwaves, droughts and floods to Europe and beyond, it is not easy putting something uplifting on paper.

Scientists have been warning for about half a century that this would happen and now climate change is a tangible reality around the world. Over the past decade reports of unusual and extreme weather events, all of which can be linked to a changing climate, have been mounting and are now a regular news item, one that we are already getting used to. Unfortunately climate change is only one of many interlinked problems we have created, reminders of which I see everywhere I go: From a walk on the beach I no longer bring shells but a collection of plastic bottles, containers and other bits and pieces of rubbish. The bird colony up the road has dwindled and the number of breeding fulmars, kittiwakes, razorbills and guillemots has about halved over the past decade. Twenty years ago fields and pastures exploded in a cacophony of colours every spring, buzzing with the sound of insects; today the only colour to be seen is green and there is an eerie quietness over the countryside.

Yet because these changes have happened gradually, few really noticed them. This phenomenon is known as Shifting Baseline Syndrome. It's like if you give a toddler a bag of building bricks. Take them away all at once, and there will be wailing; take one away every other day, and the lack of bricks will only be noticed when they are all gone. The same is happening to us. The beforementioned bird colony is a fine example. Unless someone actually counts the birds on a regular basis, a few less breeding pairs every year will hardly be noticed until one day the cliffs will stay empty and quiet.

Despite the rather dire outlook for the future, I still like to believe that we can make

Opposite: These are two of Ireland's most iconic animals: the curlew and the pine marten. The curlew is slowly disappearing as a breeding species because of a lack of habitat, which is caused by intensified farming methods. The pine marten was in the same situation not too long ago, but protection of hedgerows and – paradoxically – Sitka spruce plantations brought the little predator back from the brink of extinction, and populations are now growing. Small changes in the way Ireland is being farmed could do the same for the curlew.

changes and restore the planet to what it once was. Let the grass grow, dig a small pond, plant a hedgerow, grow your own food or pick some in the wild, try to fix things that are broken instead of throwing them away immediately — and we all should also do a bit more wailing so the rich and powerful can hear us. Wailing for the plants and animals that have disappeared because of us, wailing for the state the land and the oceans are in because of us, and wailing for what will happen to us if we don't manage to change.

We all need to become more aware of what is going on in the world around us, to look and to listen, to tune in to the natural world of which we are — like it or not — a part. Many of us are doing this already. While we have created a mess on one side, on the other we have been exploring and learning with fascinating and humbling outcomes. We now understand better than ever how the global climate works and how small changes can have big impacts. We comprehend the multitude of factors that make ecosystems work, and we are seeing proof that we are not the crowning achievement of evolution we thought we were but rather one intelligent being among many, which brings us back to the start of this book and the words of Baba Dioum: 'In the end we will conserve only what we love, we will love only what we understand, and we will understand only what we are taught.'

Carsten Krieger, Kilbaha, September 2022

Index